Quadratics Answer Book

Solutions to every question

—

Step by step :)

I got the quadratics book.

It's really good.

Expanding brackets...factoring... solving them.

Most of it makes sense.

That's good.

So what's the problem?

Sometimes I get halfway through a question and then...

I'm stuck.

Other times I finish it, but my answer's wrong.

I don't know why or where I messed up.

Quadratics isn't the easiest of topics.

There's a lot to take in.

Where are you getting stuck on?

Completing the square and factoring.

I don't get how they work.

Yeah, they can be tricky.

Is your algebra good?

Because most of the quadratics stuff needs good algebra.

Pretty good, actually.

I understand most of it.

So it's more about getting your head around quadratics.

Yeah, you could say that.

Alright, here's what you do.

When you get stuck.

Go through the worked examples again in the Quadratics book.

Then try the question you're stuck on again.

If you still can't get it.

Then find the question in this answer book.

See how it's done.

Find the step you missed or where you went wrong.

Then go back and try the question again.

See if you get it right.

If you do, great.

If you still get it wrong.

Check the steps again.

Follow each step until you get it.

Oh okay. That makes sense.

This is the fastest way you'll learn.

You don't wanna be Googling for answers or searching YouTube.

With this answer book.

It's right there.

Yep, got it.

This should help heaps.

It will.

Stuff You're About to Learn How to Answer

Expanding
Quadratics

Single Brackets

Multiple Choice

1 **Option D**

$x(2x+3)$

$= 2x^2 + 3x$

2 **Option B**

$3x(4x-7)$

$= 12x^2 - 21x$

3 **Option A**

$-4a(8-5a)$

$= -32a + 20a^2$

$= 20a^2 - 32a$

4 **Option C**

$2x(x-2) - x(x+4)$

$= 2x^2 - 4x - x^2 - 4x$

$= x^2 - 8x$

5 **Option B**

$x(5-x) - 3(x^2-1)$

$= 5x - x^2 - 3x^2 + 3$

$= -4x^2 + 5x + 3$

Time To Practice

1 a $x(x+5)$
$= x^2 + 5x$

b $y(2y+1)$
$= 2y^2 + y$

c $a(3a-1)$
$= 3a^2 - a$

d $2x(1-x)$
$= 2x - 2x^2$

e $3y(2-y)$
$= 6y - 3y^2$

f $-2a(2+3a)$
$= -4a - 6a^2$

g $-3x(6x-5)$
$= -18x^2 + 15x$

h $-5y(1-2y)$
$= -5y + 10y^2$

i $-4a(3-4a)$
$= -12a + 16a^2$

2 a $x(x+3) + x(x+4)$
$= x^2 + 3x + x^2 + 4x$
$= 2x^2 + 7x$

b $x(x-5) + x(x-7)$
$= x^2 - 5x + x^2 - 7x$
$= 2x^2 - 12x$

c $x(3x-1) + x(4-x)$
$= 3x^2 - x + 4x - x^2$
$= 2x^2 + 3x$

d $x(4x-3) - 2x(x-7)$
$= 4x^2 - 3x - 2x^2 + 14x$
$= 2x^2 + 11x$

e $x(5x+2) + 2x(6-x)$
$= 5x^2 + 2x + 12x - 2x^2$
$= 3x^2 + 14x$

f $3x(2x+4) - x(5-2x)$
$= 6x^2 + 12x - 5x + 2x^2$
$= 8x^2 + 7x$

g $4x(2x-1) + 2x(1-3x)$
$= 8x^2 - 4x + 2x - 6x^2$
$= 2x^2 - 2x$

h $2x(2-3x) - 3x(2x-7)$
$= 4x - 6x^2 - 6x^2 + 21x$
$= -12x^2 + 25x$

3 **a** $3(2x^2 - 5x + 4)$

$= 6x^2 - 15x + 12$

b $-2(x^2 - 3x - 7)$

$= -2x^2 + 6x + 14$

c $4(x - 2 - 3x^2)$

$= 4x - 8 - 12x^2$

$= -12x^2 + 4x - 8$

d $-5(9 - 8x + 6x^2)$

$= -45 + 40x - 30x^2$

$= -30x^2 + 40x - 45$

4 **a** $3x(5 - 2x) - 2(x^2 - 3)$

$= 15x - 6x^2 - 2x^2 + 6$

$= -8x^2 + 15x + 6$

b $4x(1 + 2x) - 3(x^2 + 3)$

$= 4x + 8x^2 - 3x^2 - 9$

$= 5x^2 + 4x - 9$

c $x(7 - x) - 5(x^2 - 1)$

$= 7x - x^2 - 5x^2 + 5$

$= -6x^2 + 7x + 5$

d $-2x(1 - 3x) - 4(x^2 - 6)$

$= -2x + 6x^2 - 4x^2 + 24$

$= 2x^2 - 2x + 24$

e $3(2x - 7) + 7(x^2 - x + 4)$

$= 6x - 21 + 7x^2 - 7x + 28$

$= 7x^2 - x + 7$

f $-2(4x + 3) - (-5x^2 - 3x + 4)$

$= -8x - 6 + 5x^2 + 3x - 4$

$= 5x^2 - 5x - 10$

g $6(1 + 3x - 2x^2) - 2x(3x + 1)$

$= 6 + 18x - 12x^2 - 6x^2 - 2x$

$= -18x^2 + 16x + 6$

h $-5(-x^2 + x + 3) + 2x(1 - x)$

$= 5x^2 - 5x - 15 + 2x - 2x^2$

$= 3x^2 - 3x - 15$

i $-7(6x^2 + x - 8) + 2x(1 + 3x)$

$= -42x^2 - 7x + 56 + 2x + 6x^2$

$= -36x^2 - 5x + 56$

j $-3(-4x^2 - x - 2) - 3(-x - 5)$

$= 12x^2 + 3x + 6 + 3x + 15$

$= 12x^2 + 6x + 21$

Challenge Yourself

1 $3x(2-x)-4(1-x^2)$

$3x(2-x)-4(1-x^2)$

$= 6x-3x^2-4+4x^2$

$= x^2+6x-4$

$\therefore\ x^2+6x-4 = kx^2+6x-4$

$k = 1$

2 $-4y^2+2y(3y-5)+6(y^2-y)$

$-4y^2+2y(3y-5)+6(y^2-y)$

$= -4y^2+6y^2-10y+6y^2-6y$

$= 8y^2-16y$

$\therefore\ 8y^2-16y = 8y^2-ky$

$k = 16$

3 $2(3p-p^2)+p(4p-1)-5(p+2)$

$2(3p-p^2)+p(4p-1)-5(p+2)$

$= 6\text{p}-2p^2+4p^2-p-5p-10$

$= 2p^2-10$

$\therefore\ 2p^2-10 = kp^2-10$

$k = 2$

4 $3(-x^2+x+2)+2x(x-4)$

$3(-x^2+x+2)+2x(x-4)$

$= -3x^2+3x+6+2x^2-8x$

$= -x^2-5x+6$

$\therefore\ -x^2-5x+6 = -x^2-5x+k$

$k = 6$

5 $5(2n-n^2)+3n(n-7)-4(-n^2+3n-2)$

$5(2n-n^2)+3n(n-7)-4(-n^2+3n-2)$

$=10n-5n^2+3n^2-21n+4n^2-12n+8$

$=2n^2-23n+8$

$\therefore\ 2n^2-23n+8=2n^2-kn+8$

$k=23$

6 $6t^2-3t(t+2)+4(t^2-t)+3t$

$6t^2-3t(t+2)+4(t^2-t)+3t$

$=6t^2-3t^2-6t+4t^2-4t+3t$

$=7t^2-7t$

$\therefore\ 7t^2-7t=kt^2-kt$

$k=7$

Double Brackets

Multiple Choice

1 **Option B**

$(x + 3)(x - 4)$

$= x^2 - 4x + 3x - 12$

$= x^2 - x - 12$

2 **Option C**

$(x - 2)(x - 3)$

$= x^2 - 3x - 2x + 6$

$= x^2 - 5x + 6$

3 **Option C**

$(x - 6)(2x + 3)$

$= 2x^2 + 3x - 12x - 18$

$= 2x^2 - 9x - 18$

4 **Option D**

$(4x - 1)(2x + 5)$

$= 8x^2 + 20x - 2x - 5$

$= 8x^2 + 18x - 5$

5 **Option A**

$2(x - 2)(x - 3)$

$= 2\left(x^2 - 3x - 2x + 6\right)$

$= 2\left(x^2 - 5x + 6\right)$

$= 2x^2 - 10x + 12$

Time To Practice

1 **a** $(x+3)(x+4)$
$= x^2 + 4x + 3x + 12$
$= x^2 + 7x + 12$

b $(x+2)(x+8)$
$= x^2 + 8x + 2x + 16$
$= x^2 + 10x + 16$

c $(x+8)(x+3)$
$= x^2 + 3x + 8x + 24$
$= x^2 + 11x + 24$

d $(p-2)(p+3)$
$= p^2 + 3p - 2p - 6$
$= p^2 + p - 6$

e $(p-5)(p+4)$
$= p^2 + 4p - 5p - 20$
$= p^2 - p - 20$

f $(p+3)(p-7)$
$= p^2 - 7p + 3p - 21$
$= p^2 - 4p - 21$

g $(k-4)(k-6)$
$= k^2 - 6k - 4k + 24$
$= k^2 - 10k + 24$

h $(k-5)(k-8)$
$= k^2 - 8k - 5k + 40$
$= k^2 - 13k + 40$

i $(k-1)(k-9)$
$= k^2 - 9k - k + 9$
$= k^2 - 10k + 9$

2 **a** $(2x+1)(x+5)$
$= 2x^2 + 10x + x + 5$
$= 2x^2 + 11x + 5$

b $(3x-1)(x+1)$
$= 3x^2 + 3x - x - 1$
$= 3x^2 + 2x - 1$

c $(2x-3)(x+2)$
$= 2x^2 + 4x - 3x - 6$
$= 2x^2 + x - 6$

d $(4p+5)(3p+2)$
$= 12p^2 + 8p + 15p + 10$
$= 12p^2 + 23p + 10$

2 e $(3p+2)(3p-5)$
$= 9p^2 - 15p + 6p - 10$
$= 9p^2 - 9p - 10$

f $(5p+3)(4p-2)$
$= 20p^2 - 10p + 12p - 6$
$= 20p^2 + 2p - 6$

g $(3+2k)(7-k)$
$= 21 - 3k + 14k - 2k^2$
$= -2k^2 + 11k + 21$

h $(5-2k)(3+4k)$
$= 15 + 20k - 6k - 8k^2$
$= -8k^2 + 14k + 15$

i $(11-3k)(10-7k)$
$= 110 - 77k - 30k + 21k^2$
$= 21k^2 - 107k + 110$

3 a $2(x+1)(x+3)$
$= 2(x^2 + 3x + x + 3)$
$= 2(x^2 + 4x + 3)$
$= 2x^2 + 8x + 6$

b $3(x+2)(x-7)$
$= 3(x^2 - 7x + 2x - 14)$
$= 3(x^2 - 5x - 14)$
$= 3x^2 - 15x - 42$

c $2(p-8)(p+1)$
$= 2(p^2 + p - 8p - 8)$
$= 2(p^2 - 7p - 8)$
$= 2p^2 - 14p - 16$

d $4(p-9)(p-2)$
$= 4(p^2 - 2p - 9p + 18)$
$= 4(p^2 - 11p + 18)$
$= 4p^2 - 44p + 72$

e $-5(m+3)(m+2)$
$= -5(m^2 + 2m + 3m + 6)$
$= -5(m^2 + 5m + 6)$
$= -5m^2 - 25p - 30$

f $-4(m-1)(m+6)$
$= -4(m^2 + 6m - m - 6)$
$= -4(m^2 - 5m - 6)$
$= -4m^2 + 20p + 24$

g $-2(x-3)(x-8)$
$= -2(x^2 - 8x - 3x + 24)$
$= -2(x^2 - 11x - 24)$
$= -2x^2 + 22x - 48$

h $3(2x+1)(2x+3)$
$= 3(4x^2 + 6x + 2x + 3)$
$= 3(4x^2 + 8x + 3)$
$= 12x^2 + 24x + 12$

3 **i** $6(3y-4)(y+2)$

$= 6\left(3y^2 + 6y - 4y - 8\right)$

$= 6\left(3y^2 + 2y - 8\right)$

$= 18y^2 + 12y - 48$

j $-2(z+4)(3z-7)$

$= -2\left(3z^2 - 7z + 12z - 28\right)$

$= -2\left(3z^2 + 5z - 28\right)$

$= -6z^2 - 10z + 56$

4 **a** $(x+2)(x-1) - 2x$

$= \left(x^2 - x + 2x - 2\right) - 2x$

$= x^2 + x - 2 - 2x$

$= x^2 - x - 2$

b $5x + (x+3)(x-4)$

$= 5x + \left(x^2 - 4x + 3x - 12\right)$

$= 5x + \left(x^2 - x - 12\right)$

$= 5x + x^2 - x - 12$

$= x^2 + 4x - 12$

c $3x - (2x-5)(x+2)$

$= 3x - \left(2x^2 + 4x - 5x - 10\right)$

$= 3x - \left(2x^2 - x - 10\right)$

$= 3x - 2x^2 + x + 10$

$= -2x^2 + 4x + 10$

d $2x - (3x-1)(2x-3)$

$= 2x - \left(6x^2 - 9x - 2x + 3\right)$

$= 2x - \left(6x^2 - 11x + 3\right)$

$= 2x - 6x^2 + 11x - 3$

$= -6x^2 + 13x - 3$

e $(3-7x)(2x-5) - 4x$

$= \left(6x - 15 - 14x^2 + 35x\right) - 4x$

$= \left(41x - 14x^2 - 15\right) - 4x$

$= 41x - 14x^2 + 15 - 4x$

$= -14x^2 + 37x + 15$

f $-7x - (3x-1)(2x-9)$

$= -7x - \left(6x^2 - 27x - 2x + 9\right)$

$= -7x - \left(6x^2 - 29x + 9\right)$

$= -7x - 6x^2 + 29x - 9$

$= -6x^2 + 22x - 9$

5 **a** $(x+1)(x+3) + (x+2)(x+4)$

$= \left(x^2 + 3x + x + 3\right) + \left(x^2 + 4x + 2x + 8\right)$

$= \left(x^2 + 4x + 3\right) + \left(x^2 + 6x + 8\right)$

$= x^2 + 4x + 3 + x^2 + 6x + 8$

$= 2x^2 + 10x + 11$

5 **b** $(a+3)(a-1)+(a-2)(a-4)$

$=\left(a^2-a+3a-3\right)+\left(a^2-4a-2a+8\right)$

$=\left(a^2+2a-3\right)+\left(a^2-6a+8\right)$

$=a^2+2a-3+a^2-6a+8$

$=2a^2-4a-5$

c $(2a-7)(a+4)-(a+6)(a-3)$

$=\left(2a^2+8a-7a-28\right)-\left(a^2-3a+6a-18\right)$

$=\left(2a^2+a-28\right)-\left(a^2+3a-18\right)$

$=2a^2+a-28-a^2-3a+18$

$=a^2-2a-10$

d $(y+3)(y-5)-(y+6)(2y+5)$

$=\left(y^2-5y+3y-15\right)-\left(2y^2+5y+12y+30\right)$

$=\left(y^2-2y-15\right)-\left(2y^2+17y+30\right)$

$=y^2-2y-15-2y^2-17y-30$

$=-y^2-19y-45$

e $(4b+8)(b+5)-(3b-5)(b-7)$

$=\left(4b^2+20b+8b+40\right)-\left(3b^2-21b-5b+35\right)$

$=\left(4b^2+28b+40\right)-\left(3b^2-26b+35\right)$

$=4b^2+28b+40-3b^2+26b-35$

$=b^2+54b+5$

f $(3c-5)(2c+7)-(3c+5)(c-4)$

$=\left(6c^2+21c-10c-35\right)-\left(3c^2-12c+5c-20\right)$

$=\left(6c^2+11c-35\right)-\left(3c^2-7c-20\right)$

$=6c^2+11c-35-3c^2+7c+20$

$=3c^2+18c-15$

5 **g** $(4x - 1)(3x + 5) - (7x - 9)(2x - 1)$

$= (12x^2 + 20x - 3x - 5) - (14x^2 - 7x - 18x + 9)$

$= (12x^2 + 17x - 5) - (14x^2 - 25x + 9)$

$= 12x^2 + 17x - 5 - 14x^2 + 25x - 9$

$= -2x^2 + 42x - 14$

h $(5p - 3)(p - 7) + (5p + 3)(p + 7)$

$= (5p^2 - 35p - 3p + 21) + (5p^2 + 35p + 3p + 21)$

$= (5p^2 - 38p + 21) + (5p^2 + 38p + 21)$

$= 5p^2 - 38p + 21 + 5p^2 + 38p + 21$

$= 10p^2 + 42$

i $(9 - 2y)(3 - 4y) - (3y + 2)(1 - 5y)$

$= (27 - 36y - 6y + 8y^2) - (3y - 15y^2 + 2 - 10y)$

$= (27 - 42y + 8y^2) - (-7y - 15y^2 + 2)$

$= 27 - 42y + 8y^2 + 7y + 15y^2 - 2$

$= 23y^2 - 27y + 25$

j $(6 - 5r)(3r - 2) - (-7r - 8)(1 + 4r)$

$= (18r - 12 - 15r^2 + 10r) - (-7r - 28r^2 - 8 - 32r)$

$= (28r - 12 - 15r^2) - (-39r - 28r^2 - 8)$

$= 28r - 12 - 15r^2 + 39r + 28r^2 + 8$

$= 13r^2 + 67r - 4$

k $(4w + 9)(-2w - 3) + (3w - 1)(1 - 3w)$

$= (-8w^2 - 12w - 18w - 27) + (3w - 9w^2 - 1 + 3w)$

$= (-8w^2 - 30w - 27) + (6w - 9w^2 - 1)$

$= -8w^2 - 30w - 27 + 6w - 9w^2 - 1$

$= -17w^2 - 24w - 28$

5 l $(-3q-4)(3-5q)-(5q+7)(-q-3)$

$= (-9q+15q^2-12+20q)-(-5q^2-15q-7q-21)$

$= (11q+15q^2-12)-(-5q^2-22q-21)$

$= 11q+15q^2-12+5q^2+22q+21$

$= 20q^2+33q+9$

m $-4(2a-3)(a+2)-(3a-4)(a-1)$

$= -4(2a^2+4a-3a-6)-(3a^2-3a-4a+4)$

$= -4(2a^2+a-6)-(3a^2-7a+4)$

$= -8a^2-4a+24-3a^2+7a-4$

$= -11a^2-3a+20$

n $3(5-2s)(s+2)-2(s+1)(s-2)$

$= 3(5s+10-2s^2-4s)-2(s^2-2s+s-2)$

$= 3(-2s^2+s+10)-2(s^2-s-2)$

$= -6s^2+3s+30-2s^2+2s+4$

$= -8s^2+5s+34$

o $-7(2y+5)(y+1)-2(1-4y)(y+3)$

$= -7(2y^2+2y+5y+5)-2(y+3-4y^2-12y)$

$= -7(2y^2+7y+5)-2(-4y^2-11y+3)$

$= -14y^2-49y-35+8y^2+22y-6$

$= -6y^2-27y-41$

Challenge Yourself

1 **a** $(x+4)(x+k) = x^2 + 11x + 28$

$\rightarrow 4k = 28$

$k = 7$

b $(x+7)(x-k) = x^2 + 4x - 21$

$\rightarrow -7k = -21$

$k = 3$

c $(x-k)(x-8) = x^2 - 10x + 16$

$\rightarrow 8k = 16$

$k = 2$

2 **a** $\left(x - 3\sqrt{2}\right)\left(2x - 2\sqrt{2}\right)$

$= 2x^2 - 2x\sqrt{2} - 6x\sqrt{2} + 6\left(\sqrt{2}\right)^2$

$= 2x^2 - 8x\sqrt{2} + 12$

b $\frac{1}{3}(1-3x)(2x-5)$

$= \frac{1}{3}\left(2x - 5 - 6x^2 + 15x\right)$

$= \frac{1}{3}\left(-6x^2 + 17x - 5\right)$

$= -2x^2 + \frac{17x}{3} - \frac{5}{3}$

c $\frac{(x+1)(2x-3)}{3} - \frac{(2x-1)(x+3)}{5}$

$= \frac{5\left(2x^2 - 3x + 2x - 3\right) - 3\left(2x^2 + 6x - x - 3\right)}{15}$

$= \frac{5\left(2x^2 - x - 3\right) - 3\left(2x^2 + 5x - 3\right)}{15}$

$= \frac{10x^2 - 5x - 15 - 6x^2 - 15x + 9}{15}$

$= \frac{4x^2 - 20x - 6}{15}$

Perfect Squares and DOTS

Multiple Choice

1 **Option A**

$(4m+1)^2$

$Middle\ term\ = 2(4m)(1)$

$= 8m$

2 **Option D**

$(x-3)^2$

$= x^2 - 2(x)(3) + 3^2$

$= x^2 - 6x + 9$

3 **Option B**

$(3-2p)(3-2p)$

$= 3^2 - 2(3)(2p) + (2p)^2$

$= 9 - 12p + 4p^2$

$= 4p^2 - 12p + 9$

4 **Option A**

$(5x-4)(5x+4)$

$= (5x)^2 - 4^2$

$= 25x^2 - 16$

5 **Option B**

$(1-9y)(1+9y)$

$= 1^2 - (9y)^2$

$= 1 - 81y^2$

Time To Practice

1 a $(x-4)^2$

$= x^2 - 2(x)(4) + 4^2$

$= x^2 - 8x + 16$

b $(5-y)^2$

$= 5^2 - 2(5)(y) + y^2$

$= 25 - 10y + y^2$

c $(a+2)^2$

$= a^2 + 2(a)(2) + 2^2$

$= a^2 + 4a + 4$

d $5(x+3)^2$

$= 5(x^2 + 2(x)(3) + 3^2)$

$= 5(x^2 + 6x + 9)$

$= 5x^2 + 30x + 45$

e $-2(y-4)^2$

$= -2(y^2 - 2(y)(4) + 4^2)$

$= -2(y^2 - 8y + 16)$

$= -2y^2 + 16y - 32$

f $-2(a+6)^2$

$= -2(a^2 + 2(a)(6) + 6^2)$

$= -2(a^2 + 12a + 36)$

$= -2a^2 - 24a - 72$

g $(x+7)^2 - 5x$

$= x^2 + 2(x)(7) + 7^2 - 5x$

$= x^2 + 14x + 49 - 5x$

$= x^2 + 9x + 49$

h $2(y+3)^2 - 10$

$= 2\left(y^2 + 2(y)(3) + 3^2\right) - 10$

$= 2\left(y^2 + 6y + 9\right) - 10$

$= 2y^2 + 12y + 18 - 10$

$= 2y^2 + 12y + 8$

i $-(2a+5)^2 - 10a$

$= -\left((2a)^2 + 2(2a)(5) + 5^2\right) - 10a$

$= -\left(4a^2 + 20a + 25\right) - 10a$

$= -4a^2 - 20a - 25 - 10a$

$= -4a^2 - 30a - 25$

2 **a** $(x+6)(x+6)$
$= x^2 + 2(x)(6) + 6^2$
$= x^2 + 12x + 36$

b $(y+3)(y+3)$
$= y^2 + 2(y)(3) + 3^2$
$= y^2 + 6y + 9$

c $(7-a)(7-a)$
$= 7^2 - 2(7)(a) + a^2$
$= 49 - 14a + a^2$

d $3(x+2)(x+2)$
$= 3(x^2 + 2(x)(2) + 2^2)$
$= 3(x^2 + 4x + 4)$
$= 3x^2 + 12x + 12$

e $2(y-5)(y-5)$
$= 2(y^2 - 2(y)(5) + 5^2)$
$= 2(y^2 - 10y + 25)$
$= 2y^2 - 20y + 50$

f $-6(a-2)(a-2)$
$= -6(a^2 - 2(a)(2) + 2^2)$
$= -6(a^2 - 4a + 4)$
$= -6a^2 + 24a - 24$

3 **a** $(2x-3)(2x-3) - 9$
$= (2x)^2 - 2(2x)(3) + 3^2 - 9$
$= 4x^2 - 12x + 9 - 9$
$= 4x^2 - 12x$

b $14 - (2x+5)(2x+5)$
$= 14 - [(2x)^2 + 2(2x)(5) + 5^2]$
$= 14 - (4x^2 + 20x + 25)$
$= 14 - 4x^2 - 20x - 25$
$= -4x^2 - 20x - 11$

c $5 - (3x-1)(3x-1)$
$= 5 - [(3x)^2 - 2(3x)(1) + 1^2]$
$= 5 - (9x^2 - 6x + 1)$
$= 5 - 9x^2 + 6x - 1$
$= -9x^2 + 6x + 4$

d $2 - (3-11x)(3-11x)$
$= 2 - [3^2 - 2(3)(11x) + (11x)^2]$
$= 2 - (9 - 66x + 121x^2)$
$= 2 - 9 + 66x - 121x^2$
$= -121x^2 + 66x - 7$

e $90x^2 - (2-9x)(2-9x)$
$= 90x^2 - [2^2 - 2(2)(9x) + (9x)^2]$
$= 90x^2 - (4 - 36x + 81x^2)$
$= 90x^2 - 4 + 36x - 81x^2$
$= 9x^2 + 36x - 4$

3 **f** $(3+4x)(3+4x)-2x^2$
$= 3^2 + 2(3)(4x) + (4x)^2 - 2x^2$
$= 9 + 24x + 16x^2 - 2x^2$
$= 14x^2 + 24x + 9$

4 **a** $(x+7)(x-7)$
$= x^2 - 7^2$
$= x^2 - 49$

b $(y+9)(y-9)$
$= y^2 - 9^2$
$= y^2 - 81$

c $(5-a)(5+a)$
$= 5^2 - a^2$
$= 25 - a^2$

d $(3x-1)(3x+1)$
$= (3x)^2 - 1^2$
$= 9x^2 - 1$

e $2(y-5)(y+5)$
$= 2(y^2 - 5^2)$
$= 2(y^2 - 25)$
$= 2y^2 - 50$

f $-6(a-2)(a+2)$
$= -6(a^2 - 2^2)$
$= -6(a^2 - 4)$
$= -6a^2 + 24$

5 **a** $3(2x-5)(2x+5)$
$= 3[(2x)^2 - 5^2]$
$= 3(4x^2 - 25)$
$= 12x^2 - 75$

b $2(4y-3)(4y+3)+6$
$= 2[(4y)^2 - 3^2] + 6$
$= 2(16y^2 - 9) + 6$
$= 32y^2 - 18 + 6$
$= 32y^2 - 12$

c $-(7p-3)(7p+3)+12$
$= -[(7p)^2 - 3^2] + 12$
$= -(49p^2 - 9) + 12$
$= -49p^2 + 9 + 12$
$= -49p^2 + 21$

d $5-(2-9q)(2+9q)$
$= 5 - [2^2 - (9q)^2]$
$= 5 - (4 - 81q^2)$
$= 5 - 4 + 81q^2$
$= 1 + 81q^2$

5 **e** $3x^2 - (6 - 5x)(6 + 5x)$

$= 3x^2 - [6^2 - (5x)^2]$

$= 3x^2 - (36 - 25x^2)$

$= 3x^2 - 36 + 25x^2$

$= 28x^2 - 36$

f $(4z - 1)(4z + 1) - 6z^2$

$= [(4z)^2 - 1^2] - 6z^2$

$= 16z^2 - 1 - 6z^2$

$= 10z^2 - 1$

Challenge Yourself

1 **a** $\left(x - \frac{1}{2}\right)^2$

$= x^2 - 2(x)\left(\frac{1}{2}\right) + \left(\frac{1}{2}\right)^2$

$= x^2 - x + \frac{1}{4}$

b $\left(\frac{3}{2}x + \frac{1}{3}\right)^2$

$= \left(\frac{3}{2}x\right)^2 + 2\left(\frac{3}{2}x\right)\left(\frac{1}{3}\right) + \left(\frac{1}{3}\right)^2$

$= \frac{9}{4}x^2 + x + \frac{1}{9}$

c $\left(5x - \frac{2}{3}\right)\left(5x + \frac{2}{3}\right)$

$= (5x)^2 - \left(\frac{2}{3}\right)^2$

$= 25x^2 - \frac{4}{9}$

d $2\left(\frac{3}{4}x + \frac{1}{2}\right)\left(\frac{3}{4}x - \frac{1}{2}\right)$

$= 2\left(\left(\frac{3}{4}x\right)^2 - \left(\frac{1}{2}\right)^2\right)$

$= 2\left(\frac{9}{16}x^2 - \frac{1}{4}\right)$

$= \frac{9}{8}x^2 - \frac{1}{2}$

2 **a** $4(x-2)^2-(x-1)^2$

$=4(x^2-4x+4)-(x^2-2x+1)$

$=4x^2-16x+16-x^2+2x-1$

$=3x^2-14x+15$

b $\frac{1}{3}(6-7x)(6+7x)-\frac{2}{3}x^2$

$=\frac{1}{3}(6^2-(7x^2))-\frac{2}{3}x^2$

$=\frac{1}{3}(36-49x^2)-\frac{2}{3}x^2$

$=12-\frac{49}{3}x^2-\frac{2}{3}x^2$

$=12-\frac{51}{3}x^2$

$=12-17x^2$

c $(\sqrt{2}-3x)(\sqrt{2}+3x)$

$=\left((\sqrt{2})^2-(3x)^2\right)$

$=2-9x^2$

d $(\sqrt{2}x-1)(\sqrt{2}x+1)$

$=\left((\sqrt{2}x)^2-1^2\right)$

$=2x^2-1$

3 $(x-2)^2 - 5x + k = x^2 - 9x + 13$

$Left\ Side\ = x^2 - 4x + 4 - 5x + k$

$= 4x^2 - 9x + 4 + k$

$4x^2 - 9x + 4 + k = x^2 - 9x + 13$

$\therefore 4 + k = 13$

$k = 9$

4 $(5x+1)^2 - 2(x-k) = 25x^2 + 8x + 9$

$Left\ Side\ = 25x^2 + 10x + 1 - 2x + 2k$

$= 25x^2 - 8x + 1 + 2k$

$25x^2 - 8x + 1 + 2k = 25x^2 + 8x + 9$

$\therefore\ 1 + 2k = 9$

$2k = 8$

$k = 4$

5 $(x+1)(x-1) + k(3x+1) = x^2 - 18x - 7$

$Left\ Side\ = x^2 - 1 + 3kx + k$

$= x^2 + 3kx - 1 + k$

$x^2 + 3kx - 1 + k = x^2 - 18x - 7$

$\therefore -1 + k = -7$

$k = -6$

Factoring
Quadratics

Greatest Common Factor

Multiple Choice

1 **Option C**

$6x^2 + 9$

$= 3(2x^2 + 3)$

$\therefore GCF\ is\ 3$

2 **Option B**

$15x^2 + 30$

$= 15(x^2 + 2)$

3 **Option C**

$16x^2 - 4x$

$= 4x(4x - 1)$

$\therefore GCF\ is\ 4x$

4 **Option D**

$8p - 24p^2$

$= 8p(1 - 3p)$

5 **Option D**

$18x^2 - 12x + 3$

$= 3(6x^2 - 4x + 1)$

Time To Practice

1

a $2x^2 - 4$
$= 2(x^2 - 2)$

b $3a^2 + 9$
$= 3(a^2 + 3)$

c $-12b^2 - 3$
$= -3(4b^2 + 1)$

d $15x^2 + 25$
$= 5(3x^2 + 5)$

e $27 - 9a^2$
$= 9(3 - a^2)$

f $-32b^2 - 48$
$= -16(2b^2 + 3)$

g $75x^2 - 100$
$= 25(3x^2 - 4)$

h $35a^2 - 49$
$= 7(5a^2 - 7)$

i $81b^2 - 27$
$= 27(3b^2 - 1)$

2

a $6x^2 + 8x$
$= 2x(3x + 4)$

b $4a^2 - 2a$
$= 2a(2a - 1)$

c $14b^2 - 21b$
$= 7b(2b - 3)$

d $12x - 12x^2$
$= 12x(1 - x)$

e $9a^2 - 15a$
$= 3a(3a - 5)$

f $18b - 4b^2$
$= 2b(9 - 2b)$

g $2x^2 - 32x$
$= 2x(x - 16)$

h $12a^2 - 27a$
$= 3a(4a - 9)$

i $-5b - 2b^2$
$= -(5b + 2b)$

3

a $2x^2 + 4x + 6$
$= 2(x^2 + 2x + 3)$

b $3a^2 - 6a - 12$
$= 3(a^2 - 2a - 4)$

c $12b^2 + 24b + 36$
$= 12(b^2 + 2b + 3)$

d $5x^2 - 10x + 15$
$= 5(x^2 - 2x + 3)$

e $6a^2 - 18a + 24$
$= 6(a^2 - 3a + 4)$

f $8b^2 + 8b - 32$
$= 8(b^2 + b - 4)$

g $-4x^2 - 8x - 12$
$= -4(x^2 + 2x + 3)$

h $-15a^2 + 30a - 45$
$= -15(a^2 - 2a + 3)$

i $-9b^2 + 27b - 36$
$= -9(b^2 - 3b + 4)$

4 **a** $9(x+1)+x(x+1)$
$= (x+1)(9+x)$
$= (x+1)(x+9)$

b $x(x+5)-2(x+5)$
$= (x+5)(x-2)$

c $-4(y-7)+y(y-7)$
$= (y-7)(-4+y)$ *or*
$= (y-7)(y-4)$

d $2y(y-1)-3(y-1)$
$= (y-1)(2y-3)$

e $p(3p+1)-8(3p+1)$
$= (3p+1)(p-8)$

f $-5(2p+3)-7p(2p+3)$
$= (2p+3)(-5-7p)$
$= -(2p+3)(5+7p)$ *or*
$= -(2p+3)(7p+5)$

g $2x(4x-9)-3(4x-9)$
$= (4x-9)(2x-3)$

h $-(3x-5)-4x(3x-5)$
$= (3x-5)(-1-4x)$
$= -(3x-5)(1+4x)$ *or*
$= -(3x-5)(4x+1)$

5 **a** $(x+2)^2+3(x+2)$
$= (x+2)\big((x+2)+3\big)$
$= (x+2)(x+5)$

b $(y-5)^2-4(y-5)$
$= (y-5)\big((y-5)-4\big)$
$= (y-5)(y-9)$

c $-2(x+4)^2+7(x+4)$
$= (x+4)(-2(x+4)+7)$
$= (x+4)(-2x-8+7)$
$= (x+4)(-2x-1)$
$= -(x+4)(2x+1)$

d $2(y-1)^2-9(y-1)$
$= (y-1)(2(y-1)-9)$
$= (y-1)(2y-2-9)$
$= (y-1)(2y-11)$

e $(4x-1)^2-2(4x-1)$
$= (4x-1)\big((4x-1)-2\big)$
$= (4x-1)(4x-3)$

f $(3y+1)^2+(3y+1)$
$= (3y+1)\big((3y+1)+1\big)$
$= (3y+1)(3y+2)$

g $(3-2x)^2+2(3-2x)$

$=(3-2x)\big((3-2x)+2\big)$

$=(3-2x)(5-2x)$ *or*

$=-(2x-3)(5-2x)$ *or*

$=-(3-2x)(2x-5)$

h $-4(2y+5)^2+3(2y+5)$

$=(2y+5)(-4(2y+5)+3)$

$=(2y+5)(-8y-20+3)$

$=(2y+5)(-8y-17)$ *or*

$=-(2y+5)(8y+17)$

Challenge Yourself

1 **a** a^2b-ab^2

$=ab(a-b)$

b $2x^2yz-4xy$

$=2xy(xz-2)$

c $-12m^2n-12mn^2$

$=-12mn(m+n)$

d $6xyz^2-3yz^2$

$=3yz^2(2x-1)$

e $-2p^2q+4pq-6p^2$

$=-2p(pq-2q+3p)$

f $5a^2b-10ab^2-15ab$

$=5ab(a-2b-3)$

g $3(x-1)(x+2)+2(x-1)$

$=(x-1)(3(x+2)+2)$

$=(x-1)(3x+6+2)$

$=(x-1)(3x+8)$

h $5(y-2)-10(2y-1)(y-2)$

$=5(y-2)\big(1-2(2y-1)\big)$

$=5(y-2)(1-4y+2)$

$=5(y-2)(3-4y)$ *or*

$=-5(y-2)(4y-3)$

Perfect Squares

Multiple Choice

1 **Option B**

$x^2 + 6x + 9$

$= x^2 + 6x + 3^2$

$= (x + 3)(x + 3)$

2 **Option A**

$x^2 - 16x + 64$

$= x^2 - 16x + 8^2$

$= (x - 4)(x - 4)$

3 **Option C**

$-x^2 - 12x - 36$

$= -(x^2 + 12x + 6^2)$

$= -(x + 6)^2$

4 **Option B**

$4x^2 - 20x + 25$

$= (2x)^2 - 20x + 5^2$

$= (2x - 5)^2$

5 **Option D**

$16 - 48x + 36x^2$

$= 4(4 - 12x + 9x^2)$

$= 4(2^2 - 12x + (3x)^2)$

$= 4(2 - 3x)^2$

Time To Practice

1 **a** $x^2 + 14x + 49$
$= x^2 + 14x + 7^2$
$= (x + 7)^2$

b $a^2 - 6a + 9$
$= a^2 - 6x + 3^2$
$= (a - 3)^2$

c $b^2 + 2b + 1$
$= b^2 + 2b + 1^2$
$= (b + 1)^2$

d $x^2 - 4x + 4$
$= x^2 - 4x + 2^2$
$= (x - 2)^2$

e $-a^2 - 18a - 81$
$= -(a^2 + 18a + 9^2)$
$= -(a + 9)^2$

f $25 - 10b + b^2$
$= 5^2 - 10b + b^2$
$= (5 - b)^2$

g $-x^2 - 16x - 64$
$= -(x^2 + 16x + 8^2)$
$= -(x + 8)^2$

h $-a^2 + 22a - 121$
$= -(a^2 - 22a + 11^2)$
$= -(a - 11)^2$

i $-144 + 24b - b^2$
$= -(12^2 - 24b + b^2)$
$= -(12 - b)^2$

2 **a** $9x^2 + 12x + 4$
$= (3x)^2 + 12x + 2^2$
$= (3x + 2)^2$

b $16a^2 - 40a + 25$
$= (4a)^2 - 40a + 5^2$
$= (4a - 5)^2$

c $-25b^2 + 30b - 9$
$= -[(5b)^2 - 30b + 3^2]$
$= -(5b - 3)^2$

d $-49x^2 - 28x - 4$
$= -[(7x)^2 + 28x + 2^2]$
$= -(7x + 2)^2$

e $36a^2 + 84a + 49$
$= [(6a)^2 + 84a + 7^2]$
$= (6a + 7)^2$

f $-81b^2 + 90b - 25$
$= -[(9b)^2 - 90b + 5^2]$
$= -(9b - 5)^2$

2 **g** $100x^2 - 60x + 9$
$= (10x)^2 - 60x + 3^2$
$= (10x - 3)^2$

h $-4a^2 - 20a - 25$
$= -[(2a)^2 + 20a + 5^2]$
$= -(2a + 5)^2$

i $1 + 22b + 121b^2$
$= 1^2 + 22b + (11b)^2$
$= (1 + 11b)^2$

3 **a** $3x^2 + 30x + 75$
$= 3(x^2 + 10x + 25)$
$= 3(x^2 + 10x + 5^2)$
$= 3(x + 5)^2$

b $3a^2 - 24a + 48$
$= 3(a^2 - 8a + 16)$
$= 3(a^2 - 8a + 4^2)$
$= 3(a - 4)^2$

c $2b^2 - 16b + 32$
$= 2(b^2 - 8b + 16)$
$= 2(b^2 - 8b + 4^2)$
$= 2(b - 4)^2$

d $4x^2 + 24x + 36$
$= 4(x^2 + 6x + 9)$
$= 4(x^2 + 6x + 3^2)$
$= 4(x + 3)^2$

e $-9a^2 + 54a - 81$
$= -9(a^2 - 6a + 9)$
$= -9(a^2 - 6a + 3^2)$
$= -9(a - 3)^2$

f $100 - 100b + 25b^2$
$= 25(4 - 4b + b^2)$
$= 25(2^2 - 4a + b^2)$
$= 25(2 - b)^2$

g $-16x^2 - 32x - 16$
$= -16(x^2 + 2x + 1)$
$= -16(x^2 + 2x + 1^2)$
$= -16(x + 1)^2$

h $10a^2 - 60a + 90$
$= 10(a^2 - 6a + 9)$
$= 10(a^2 - 6a + 3^2)$
$= 10(a - 3)^2$

i $-50 + 140b - 98b^2$
$= -2(25 - 70b + 49b^2)$
$= -2(5^2 - 70b + (7b)^2)$
$= -2(5 - 7b)^2$

Challenge Yourself

1 a $(x+5)^2-12(x+5)+36$
$=(x+5)^2-12(x+5)+6^2$
$=(x+5-6)^2$
$=(x-1)^2$

b $(a+b)^2+12(a+b)+36$
$=(a+b)^2+12(a+b)+6^2$
$=(a+b+6)^2$

c x^4+10x^2+25
$=(x^2)^2+10x^2+5^2$
$=(x^2+5)^2$

d $4a^8-28a^4+49$
$=(2x^4)^2-28x^4+7^2$
$=(2x^4-7)^2$

e $-16x^6+56x^3-49$
$=-(16x^6-56x^3+49)$
$=-\left[(4x^3)^2-56x^3+7^2\right]$
$=-(4x^3-7)^2$

f $9x^4-12x^2y^2+4y^4$
$=(3x^2)^2-12x^2y^2+(2y^2)^2$
$=(3x^2-2y^2)^2$

Difference of Two Squares

Multiple Choice

1 **Option C**

$x^2 - 16$

$= x^2 - 4^2$

$= (x - 4)(x + 4)$

2 **Option B**

$25 - y^2$

$= 5^2 - y^2$

$= (5 - y)(5 + y)$

3 **Option A**

$36a^2 - 49$

$= (6a)^2 - 7^2$

$= (6a - 7)(6a + 7)$

4 **Option D**

$4m^2 - 100$

$= 4(m^2 - 25)$

$= 4(m^2 - 5^2)$

$= 4(m - 5)(m + 5)$

5 **Option A**

$81 - 9x^2$

$= -9(x^2 - 9)$

$= -9(x^2 - 3^2)$

$= -9(x - 3)(x + 3)$

Time To Practice

1 a $x^2 - 1$

$= x^2 - 1^2$

$= (x - 1)(x + 1)$

b $a^2 - 9$

$= a^2 - 3^2$

$= (a - 3)(a + 3)$

c $b^2 - 121$

$= b^2 - 11^2$

$= (b - 11)(b + 11)$

d $144 - x^2$

$= 12^2 - x^2$

$= (12 - x)(12 + x)$

e $16 - a^2$

$= 4^2 - a^2$

$= (4 - a)(4 + a)$

f $1 - 100b^2$

$= 1^2 - (10b)^2$

$= (1 - 10b)(1 + 10b)$

g $36x^2 - 1$

$= (6x)^2 - 1^2$

$= (6x - 1)(6x + 1)$

h $4a^2 - 25$

$= (2a)^2 - 5^2$

$= (2a - 5)(2a + 5)$

i $81b^2 - 16$

$= (9b)^2 - 4^2$

$= (9b - 4)(9b + 4)$

j $49 - 9x^2$

$= 7^2 - (3x)^2$

$= (7 - 3x)(7 + 3x)$

k $64 - 25a^2$

$= 8^2 - (5a)^2$

$= (8 - 5a)(8 + 5a)$

l $36b^2 - 1$

$= (6b)^2 - 1^2$

$= (6b - 1)(6b + 1)$

m $12x^2 - 48$

$= 12(x^2 - 4)$

$= 12(x^2 - 2^2)$

$= 12(x - 2)(x + 2)$

n $44 - 11a^2$

$= -11(a^2 - 4)$

$= -11(a^2 - 2^2)$

$= -11(a - 2)(a + 2)$

1 o $3b^2 - 27$
$= 3(b^2 - 9)$
$= 3(b^2 - 3^2)$
$= 3(b - 3)(b + 3)$

p $75 - 3x^2$
$= -3(x^2 - 25)$
$= -3(x^2 - 5^2)$
$= -3(x - 5)(x + 5)$

q $50a^2 - 8$
$= 2(25a^2 - 4)$
$= 2[(5a)^2 - 2^2]$
$= 2(5a - 2)(5a + 2)$

r $54 - 6b^2$
$= -6(b^2 - 9)$
$= -6(b^2 - 3^2)$
$= -6(b - 3)(b + 3)$

s $72x^2 - 98$
$= 2(36x^2 - 49)$
$= 2[(6x)^2 - 7^2]$
$= 2(6x - 7)(6x + 7)$

t $-27a^2 + 12$
$= -3(9a^2 - 4)$
$= -3[(3a)^2 - 2^2]$
$= -3(3a - 2)(3a + 2)$

u $125 - 20b^2$
$= -5(4b^2 - 25)$
$= -5[(2b)^2 - 5^2]$
$= -5(2b - 5)(2b + 5)$

2 a $(x + 2)^2 - 4$
$= (x + 2)^2 - 2^2$
$= (x + 2 - 2)(x + 2 + 2)$
$= x(x + 4)$

b $(a + 5)^2 - 49$
$= (a + 5)^2 - 7^2$
$= (a + 5 - 7)(a + 5 + 7)$
$= (a - 2)(a + 12)$

c $(b - 2)^2 - 1$
$= (b - 2)^2 - 1^2$
$= (b - 2 - 1)(b - 2 + 1)$
$= (b - 3)(b - 1)$

d $(2x - 1)^2 - 1$
$= (2x - 1)^2 - 1^2$
$= (2x - 1 - 1)(2x - 1 + 1)$
$= (2x - 2)2x$
$= 4x(x - 1)$

2 **e** $(4a-5)^2-25$

$=(4a-5)^2-5^2$

$=(4a-5-5)(4a-5+5)$

$=(4a-10)4a$

$=8a(2a-5)$

f $(5b+1)^2-64$

$=(5b+1)^2-8^2$

$=(5b+1-8)(5b+1+8)$

$=(5b-7)(5b+9)$

g $4-(2x-3)^2$

$=2^2-(2x-3)^2$

$=(2-(2x-3))(2+(2x-3))$

$=(2-2x+3)(2+2x-3)$

$=(5-2x)(-1+2x)$

$=(5-2x)(2x-1)$

h $81-(7a+1)^2$

$=9^2-(7a+1)^2$

$=(9-(7a+1))(9+(7a+1))$

$=(9-7a-1)(9+7a+1)$

$=(8-7a)(7a+10)$

i $9-(3b-2)^2$

$=3^2-(3b-2)^2$

$=(3-(3b-2))(3+(3b-2))$

$=(3-3b+2)(3+3b-2)$

$=(5-3b)(3b+1)$

j $2(x+5)^2-18$

$=2[(x+5)^2-9]$

$=2[(x+5)^2-3^2]$

$=2(x+5-3)(x+5+3)$

$=2(x+2)(x+8)$

k $-3(a-2)^2+27$

$=-3[(a-2)^2-9]$

$=-3[(a-2)^2-3^2]$

$=-3(a-2-3)(a-2+3)$

$=-3(a-5)(a+1)$

l $2(1-5b)^2-98$

$=2[(1-5b)^2-49]$

$=2[(1-5b)^2-7^2]$

$=2(1-5b-7)(1-5b+7)$

$=2(-6-5b)(8-5b)$

$=-2(6+5b)(8-5b)$

3 **a** x^2-7

$=x^2-\left(\sqrt{7}\right)^2$

$=(x-\sqrt{7})(x+\sqrt{7})$

b a^2-3

$=a^2-\left(\sqrt{3}\right)^2$

$=\left(a-\sqrt{3}\right)\left(a+\sqrt{3}\right)$

3 **c** $5 - b^2$

$= \left(\sqrt{5}\right)^2 - b^2$

$= \left(\sqrt{5} - b\right)\left(\sqrt{5} + b\right)$

d $4x^2 - 13$

$= (2x)^2 - \left(\sqrt{13}\right)^2$

$= \left(2x - \sqrt{13}\right)\left(2x + \sqrt{13}\right)$

e $9a^2 - 19$

$= (3a)^2 - \left(\sqrt{19}\right)^2$

$= \left(3a - \sqrt{19}\right)\left(3a + \sqrt{19}\right)$

f $3b^2 - 66$

$= 3\left(b^2 - 22\right)$

$= 3\left(b^2 - \left(\sqrt{22}\right)^2\right)$

$= 3\left(b - \sqrt{22}\right)\left(b + \sqrt{22}\right)$

g $15 - 5x^2$

$= -5\left(x^2 - 3\right)$

$= -5\left(x^2 - \left(\sqrt{3}\right)^2\right)$

$= -5\left(x - \sqrt{3}\right)\left(x + \sqrt{3}\right)$

h $4 - 2a^2$

$= -2\left(a^2 - 2\right)$

$= -2\left(a^2 - \left(\sqrt{2}\right)^2\right)$

$= -2\left(a - \sqrt{2}\right)\left(a + \sqrt{2}\right)$

i $36 - 48b^2$

$= -12\left(4b^2 - 3\right)$

$= -12\left((2b)^2 - \left(\sqrt{3}\right)^2\right)$

$= -12\left(2b - \sqrt{3}\right)\left(2b + \sqrt{3}\right)$

j $(x + 3)^2 - 6$

$= (x + 3)^2 - \left(\sqrt{6}\right)^2$

$= \left(x + 3 - \sqrt{6}\right)\left(x + 3 - \sqrt{6}\right)$

3 **k** $15-(a-4)^2$

$=\left(\sqrt{15}\right)^2-(a-4)^2$

$=\left(\sqrt{15}-(a-4)\right)\left(\sqrt{15}+(a-4)\right)$

$=\left(\sqrt{15}-a+4\right)\left(\sqrt{15}+a-4\right)$

$=\left(\sqrt{15}+4-a\right)\left(\sqrt{15}-4+a\right)$

or

$=-\left(a-4-\sqrt{15}\right)\left(a-4+\sqrt{15}\right)$

l $-2(b-13)^2+14$

$=-2[(b-13)^2-7]$

$=-2\left((b-13)^2-\sqrt{7}\right)\left((b-13)^2+\sqrt{7}\right)$

$=-2\left(b-13-\sqrt{7}\right)\left(b-13+\sqrt{7}\right)$

Challenge Yourself

1 a $3ax^2 - 147a$
$= 3a(x^2 - 49)$
$= 3a(x^2 - 7^2)$
$= 3a(x - 7)(x + 7)$

b $4m^2p - 36n^2p$
$= 4p(m^2 - 9n^2)$
$= 4p[m^2 - (3n)^2]$
$= 4p(m - 3n)(m + 3n)$

c $9x^2 - y^2z^2$
$= (3x)^2 - (yz)^2$
$= (3x - yz)(3x + yz)$

d $4x^2 - \frac{y^2}{121}$
$= (2x)^2 - \left(\frac{y}{11}\right)^2$
$= \left(2x - \frac{y}{11}\right)\left(2x + \frac{y}{11}\right)$

e $\frac{x^2}{4} - \frac{9}{25}$
$= \left(\frac{x}{2}\right)^2 - \left(\frac{3}{5}\right)^2$
$= \left(\frac{x}{2} - \frac{3}{5}\right)\left(\frac{x}{2} + \frac{3}{5}\right)$

f $\frac{5x^2}{2} - \frac{5}{2}$
$= \frac{5}{2}(x^2 - 1)$
$= \frac{5}{2}(x - 1)(x + 1)$

g $\frac{a^2b^2}{36} - \frac{c^2}{16}$
$= \left(\frac{ab}{6}\right)^2 - \left(\frac{c}{4}\right)^2$
$= \left(\frac{ab}{6} - \frac{c}{4}\right)\left(\frac{ab}{6} + \frac{c}{4}\right)$

h $\frac{p^2q^2}{81} - \frac{m^2n^2}{100}$
$= \left(\frac{pq}{9}\right)^2 - \left(\frac{mn}{10}\right)^2$
$= \left(\frac{pq}{9} - \frac{mn}{10}\right)\left(\frac{pq}{9} + \frac{mn}{10}\right)$

The AC Method

Multiple Choice

1 **Option D**

First step, multiply the a and c values.

2 **Option A**

$3x^2 + 4x - 4$

$AC = 3 \times -4$

$= -12$

3 **Option C**

$a^2 - 7a + 10$

$= (a - 5)(a - 2)$

$ac = 10, b = -7$

$-5 \times -2 = 10 \ \rightarrow ac$
$-5 + (-2) = -7 \ \rightarrow b$

4 **Option B**

$3p^2 - 8p - 3$

$ac = -9, b = -8$

$-9 \times 1 = -9 \ \rightarrow ac$
$-9 + 1 = -8 \ \rightarrow b$

$\therefore -9 \ and \ 1$

5 **Option C**

$2m^2 + 5m + 3$

$= (m + 1)(2m + 3)$

$ac = 6, b = 5$

$2 \times 3 = 6 \ \rightarrow ac$
$2 + 3 = 5 \ \rightarrow b$

$\frac{2}{2} = \frac{1}{1} \ and \ \frac{3}{2}$

Time To Practice

1 **a** $x^2 + 4x + 3$

$= (x + 1)(x + 3)$

$ac = 3, b = 4$

$\mathbf{1 \times 3 = 3} \rightarrow \boldsymbol{ac}$
$\mathbf{1 + 3 = 4} \rightarrow \boldsymbol{b}$

b $a^2 - 7a + 10$

$= (a - 5)(a - 2)$

$ac = 10, b = -7$

$\mathbf{-5 \times -2 = 10} \rightarrow \boldsymbol{ac}$
$\mathbf{-5 + (-2) = -7} \rightarrow \boldsymbol{b}$

c $b^2 - 4b - 12$

$= (b - 6)(b + 2)$

$ac = -12, b = -4$

$\mathbf{-6 \times 2 = -12} \rightarrow \boldsymbol{ac}$
$\mathbf{-6 + 2 = -4} \rightarrow \boldsymbol{b}$

d $x^2 - 8x + 15$

$= (x - 5)(x - 3)$

$ac = 15, b = -8$

$\mathbf{-5 \times -3 = 15} \rightarrow \boldsymbol{ac}$
$\mathbf{-5 + (-3) = -8} \rightarrow \boldsymbol{b}$

e $a^2 - 9a + 20$

$= (a - 5)(a - 4)$

$ac = 20, b = -9$

$\mathbf{-5 \times -4 = 20} \rightarrow \boldsymbol{ac}$
$\mathbf{-5 + (-4) = -9} \rightarrow \boldsymbol{b}$

f $b^2 + b - 110$

$= (b - 10)(b + 11)$

$ac = -110, b = 1$

$\mathbf{-10 \times 11 = -110} \rightarrow \boldsymbol{ac}$
$\mathbf{-10 + 11 = 1} \rightarrow \boldsymbol{b}$

g $-x^2 + 6x - 8$

$= -(x^2 - 6x + 8)$

$= -(x - 4)(x - 2)$

$ac = 8, b = -6$

$\mathbf{-4 \times -2 = 8} \rightarrow \boldsymbol{ac}$
$\mathbf{-4 + (-2) = -6} \rightarrow \boldsymbol{b}$

h $-a^2 - 3a + 10$

$= -(a^2 + 3a - 10)$

$= -(a - 2)(a + 5)$

$ac = -10, b = 3$

$\mathbf{-2 \times 5 = -10} \rightarrow \boldsymbol{ac}$
$\mathbf{-2 + 5 = 3} \rightarrow \boldsymbol{b}$

1 i $-b^2 - 5b + 24$

$= -(b^2 + 5b - 24)$

$= -(b-3)(b+8)$

$ac = -24, b = 5$

$\mathbf{-3 \times 8 = 24} \rightarrow \boldsymbol{ac}$

$\mathbf{-3 + 8 = 5} \rightarrow \boldsymbol{b}$

j $4x^2 + 20x + 16$

$= 4(x^2 + 5x + 4)$

$= 4(x+1)(x+4)$

$ac = 4, b = 5$

$\mathbf{1 \times 4 = 4} \rightarrow \boldsymbol{ac}$

$\mathbf{1 + 4 = 5} \rightarrow \boldsymbol{b}$

k $6a^2 - 12a - 18$

$= 6(a^2 - 2a - 3)$

$= 6(a-3)(a+1)$

$ac = -3, b = -2$

$\mathbf{-3 \times 1 = -3} \rightarrow \boldsymbol{ac}$

$\mathbf{-3 + 1 = -2} \rightarrow \boldsymbol{b}$

l $2b^2 - 26b + 84$

$= 2(b^2 - 13b + 42)$

$= 2(b-7)(b-6)$

$ac = 42, b = -13$

$\mathbf{-7 \times -6 = 42} \rightarrow \boldsymbol{ac}$

$\mathbf{-7 + (-6) = -13} \rightarrow \boldsymbol{b}$

2 a $12 + 18x + 6x^2$

$= 6x^2 + 18x + 12$

$= 6(x^2 + 3x + 2)$

$= 6(x+1)(x+2)$

$ac = 2, b = 3$

$\mathbf{1 \times 2 = 2} \rightarrow \boldsymbol{ac}$

$\mathbf{1 + 2 = 3} \rightarrow \boldsymbol{b}$

b $18 - 3a^2 - 15a$

$= -3a^2 - 15a + 18$

$= -3(a^2 + 5a - 6)$

$= -3(a-1)(a+6)$

$ac = -6, b = 5$

$\mathbf{-1 \times 6 = -6} \rightarrow \boldsymbol{ac}$

$\mathbf{-1 + 6 = 5} \rightarrow \boldsymbol{b}$

c $14b - 24 - 2b^2$

$= -2b^2 + 14b - 24$

$= -2(b^2 - 7b + 12)$

$= -2(b-4)(b-3)$

$ac = 12, b = -7$

$\mathbf{-4 \times -3 = 12} \rightarrow \boldsymbol{ac}$

$\mathbf{-4 + (-3) = -7} \rightarrow \boldsymbol{b}$

2 **d** $9x^2 - 27x + 18$

$= 9(x^2 - 3x + 2)$

$= 9(x - 2)(x - 1)$

$ac = 2, b = -3$

$\mathbf{-2 \times -1 = 2} \rightarrow \boldsymbol{ac}$
$\mathbf{-2 + (-1) = -3} \rightarrow \boldsymbol{b}$

e $-5a^2 - 25a - 30$

$= -5(a^2 + 5a + 6)$

$= -5(a + 2)(a + 3)$

$ac = 6, b = 5$

$\mathbf{2 \times 3 = 6} \rightarrow \boldsymbol{ac}$
$\mathbf{2 + 3 = 5} \rightarrow \boldsymbol{b}$

f $28b - 48 - 4b^2$

$= -4b^2 + 28b - 48$

$- -4(b^2 - 7b + 12)$

$= -4(b - 4)(b - 3)$

$ac = 12, b = -7$

$\mathbf{-4 \times -3 = 12} \rightarrow \boldsymbol{ac}$
$\mathbf{-4 + (-3) = -7} \rightarrow \boldsymbol{b}$

g $15x^2 + 45x + 30$

$= 15(x^2 + 3x + 2)$

$= 15(x + 1)(x + 2)$

$ac = 2, b = 3$

$\mathbf{1 \times 2 = 2} \rightarrow \boldsymbol{ac}$
$\mathbf{1 + 2 = 3} \rightarrow \boldsymbol{b}$

h $27a - 18 - 9a^2$

$= -9a^2 + 27a - 18$

$= -9(a^2 - 3a + 2)$

$= -9(a - 2)(a - 1)$

$ac = 2, b = -3$

$\mathbf{-2 \times -1 = 2} \rightarrow \boldsymbol{ac}$
$\mathbf{-2 + (-1) = -3} \rightarrow \boldsymbol{b}$

i $48b - 6b^2 - 42$

$= -6b^2 + 48b - 42$

$= -6(b^2 - 8b + 7)$

$= -6(b - 7)(b - 1)$

$ac = 7, b = -8$

$\mathbf{-7 \times -1 = 7} \rightarrow \boldsymbol{ac}$
$\mathbf{-7 + (-1) = -8} \rightarrow \boldsymbol{b}$

3 **a** $2x^2 + 5x + 3$

$= (x+1)(2x+3)$

$ac = 6, b = 5$

$2 \times 3 = 6 \rightarrow ac$
$2 + 3 = 5 \rightarrow b$

$\mathbf{\frac{2}{2} = \frac{1}{1}}$ ***and*** $\mathbf{\frac{3}{2}}$

b $3a^2 - 8a + 4$

$= (a-2)(3a-2)$

$ac = 12, b = -8$

$-6 \times -2 = 12 \rightarrow ac$
$-6 + (-2) = -8 \rightarrow b$

$\mathbf{\frac{-6}{3} = \frac{-2}{1}}$ ***and*** $\mathbf{\frac{-2}{3}}$

c $4b^2 + 4b - 3$

$= (2b-1)(2b+3)$

$ac = -12, b = 4$

$-2 \times 6 = -12 \rightarrow ac$
$-2 + 6 = 4 \rightarrow b$

$\mathbf{\frac{-2}{4} = \frac{-1}{2}}$ ***and*** $\mathbf{\frac{6}{4} = \frac{3}{2}}$

d $6x^2 + 17x + 5$

$= (3x+1)(2x+5)$

$ac = 30, b = 17$

$2 \times 15 = 30 \rightarrow ac$
$2 + 15 = 17 \rightarrow b$

$\mathbf{\frac{2}{6} = \frac{1}{3}}$ ***and*** $\mathbf{\frac{15}{6} = \frac{5}{2}}$

e $6a^2 + 7a - 5$

$= (2a-1)(3a+5)$

$ac = -30, b = 7$

$-3 \times 10 = -30 \rightarrow ac$
$-3 + 10 = 7 \rightarrow b$

$\mathbf{\frac{-3}{6} = \frac{-1}{2}}$ ***and*** $\mathbf{\frac{10}{6} = \frac{5}{3}}$

3 **f** $8b^2 + 14b + 3$

$= (4b + 1)(2b + 3)$

$ac = 24, b = 14$

$2 \times 12 = 24 \rightarrow ac$
$2 + 12 = 14 \rightarrow b$

$\mathbf{\frac{2}{8} = \frac{1}{4}}$ ***and*** $\mathbf{\frac{12}{8} = \frac{3}{2}}$

g $12x^2 + 5x - 3$

$= (3x - 1)(4x + 3)$

$ac = -36, b = 5$

$-4 \times 9 = 36 \rightarrow ac$
$-4 + 9 = 5 \rightarrow b$

$\mathbf{\frac{-4}{12} = \frac{-1}{3}}$ ***and*** $\mathbf{\frac{9}{12} = \frac{3}{4}}$

h $-9a^2 + 3a + 2$

$= -(9a^2 - 3a - 2)$

$= -(3a - 2)(3a + 1)$

$ac = -18, b = -3$

$-6 \times 3 = -18 \rightarrow ac$
$-6 + 3 = -3 \rightarrow b$

$\mathbf{\frac{-6}{9} = \frac{-2}{3}}$ ***and*** $\mathbf{\frac{3}{9} = \frac{1}{3}}$

i $-6b^2 + 13b + 5$

$= -(6b^2 - 13b - 5)$

$= -(2b - 5)(3b + 1)$

$ac = -30, b = -13$

$-15 \times 2 = -30 \rightarrow ac$
$-15 + 2 = -13 \rightarrow b$

$\mathbf{\frac{-15}{6} = \frac{-5}{2}}$ ***and*** $\mathbf{\frac{2}{6} = \frac{1}{3}}$

j $-3x^2 + 11x - 10$

$= -(3x^2 - 11x + 10)$

$= -(x - 2)(3x - 5)$

$ac = 30, b = -11$

$-6 \times -5 = -30 \rightarrow ac$
$-6 + (-5) \rightarrow b$

$\mathbf{\frac{-6}{3} = \frac{-2}{1}}$ ***and*** $\mathbf{\frac{-5}{3}}$

3 **k** $-2a^2 - 15a - 18$

$= -(2a^2 + 15a + 18)$

$= -(2a + 3)(a + 6)$

$ac = 36, b = 15$

$3 \times 12 = 36 \rightarrow ac$
$3 + 12 = 15 \rightarrow b$

$\frac{3}{2}$ ***and*** $\frac{12}{2} = \frac{6}{1}$

l $-4b^2 + 7b + 2$

$= -(4b^2 - 7b - 2)$

$= -(b - 2)(4b + 1)$

$ac = -8, b = -7$

$-8 \times 1 = -8 \rightarrow ac$
$-8 + 1 = -7 \rightarrow b$

$\frac{-8}{4} = \frac{-2}{1}$ ***and*** $\frac{1}{4}$

4 **a** $6x^2 - 10x - 4$

$= 2(3x^2 - 5x - 2)$

$= 2(x - 2)(3x + 1)$

$ac = -6, b = -5$

$-6 \times 1 = -6 \rightarrow ac$
$-6 + 1 = -5 \rightarrow b$

$\frac{-6}{3} = \frac{-2}{1}$ ***and*** $\frac{1}{3}$

b $-6a^2 - 9a + 6$

$= -3(2a^2 + 3a - 2)$

$= -3(2a - 1)(a + 2)$

$ac = -4, b = 3$

$-1 \times 4 = -4 \rightarrow ac$
$-1 + 4 = 3 \rightarrow b$

$\frac{-1}{2} =$ ***and*** $\frac{4}{2} = \frac{2}{1}$

c $8b^2 - 4b - 12$

$= 4(2b^2 - b - 3)$

$= 4(2b - 3)(b + 1)$

$ac = -6, b = -1$

$-3 \times 2 = -6 \rightarrow ac$
$-3 + 2 = -1 \rightarrow b$

$\frac{-3}{2}$ ***and*** $\frac{2}{2} = \frac{1}{1}$

4 **d** $-15x^2 - 25x + 10$

$= -5(3x^2 + 5x - 2)$

$= -5(3x - 1)(x + 2)$

$ac = -6, b = 5$

$-1 \times 6 = -6 \rightarrow ac$
$-1 + 6 = 5 \rightarrow b$

$\frac{-1}{3}$ ***and*** $\frac{6}{3} = \frac{2}{1}$

e $14a^2 + 49a + 21$

$= 7(2a^2 + 7a + 3)$

$= 7(2a + 1)(a + 3)$

$ac = 6, b = 7$

$1 \times 6 = 6 \rightarrow ac$
$1 + 6 = 7 \rightarrow b$

$\frac{1}{2}$ ***and*** $\frac{6}{2} = \frac{3}{1}$

f $-12b^2 + 54b - 60$

$= -6(2b^2 - 9b + 10)$

$= -6(2b - 5)(b - 2)$

$ac = 20, b = -9$

$-5 \times -4 = 20 \rightarrow ac$
$-5 + (-4) = -9 \rightarrow b$

$\frac{-5}{2}$ ***and*** $\frac{-4}{2} = \frac{-2}{1}$

g $44x^2 - 132x - 77$

$= 11(4x^2 - 12x - 7)$

$= 11(2x - 7)(2x + 1)$

$ac = -28, b = -12$

$-14 \times 2 = -28 \rightarrow ac$
$-14 + 2 = -12 \rightarrow b$

$\frac{-14}{4} = \frac{-7}{2}$ ***and*** $\frac{2}{4} = \frac{1}{2}$

h $30a^2 + 25a - 30$

$= 5(6a^2 + 5a - 6)$

$= 5(3a - 2)(2a + 3)$

$ac = -36, b = 5$

$-4 \times 9 = -36 \rightarrow ac$
$-4 + 9 = 5 \rightarrow b$

$\frac{-4}{6} = \frac{-2}{3}$ ***and*** $\frac{9}{6} = \frac{3}{2}$

4 i $-16b^2 + 20b - 6$

$= -2(8b^2 - 10b + 3)$

$= -2(4b - 3)(2b - 1)$

$ac = 24, b = -10$

$-6 \times -4 = 24 \rightarrow ac$
$-6 + (-4) = -10 \rightarrow b$

$\frac{-6}{8} = \frac{-3}{4}$ ***and*** $\frac{-4}{8} = \frac{-1}{2}$

5 a $12x + 18x^2 - 30$

$= 18x^2 + 12x - 30$

$= 6(3x^2 + 2x - 5)$

$= 6(x - 1)(3x + 5)$

$ac = -15, b = 2$

$-3 \times 5 = -15 \rightarrow ac$
$-3 + 5 = 2 \rightarrow b$

$\frac{-3}{3} = \frac{-1}{1}$ ***and*** $\frac{5}{3}$

b $20 - 6a^2 + 2a$

$= -6a^2 + 2a + 20$

$= -2(3a^2 - a - 10)$

$= -2(a - 2)(3a + 5)$

$ac = -30, b = -1$

$-6 \times 5 = -30 \rightarrow ac$
$-6 + 5 = -1 \rightarrow b$

$\frac{-6}{3} = \frac{-2}{1}$ ***and*** $\frac{5}{3}$

c $48 - 20b - 8b^2$

$= -8b^2 - 20b + 48$

$= -4(2b^2 + 5b - 12)$

$= -4(2b - 3)(b + 4)$

$ac = -24, b = 5$

$-3 \times 8 = -24 \rightarrow ac$
$-3 + 8 = 5 \rightarrow b$

$\frac{-3}{2}$ ***and*** $\frac{8}{2} = \frac{4}{1}$

d $21x^2 - 21 - 56x$

$= 21x^2 - 56x - 21$

$= 7(3x^2 - 8x - 3)$

$= 7(x - 3)(3x + 1)$

$ac = -9, b = -8$

$-9 \times 1 = -9 \rightarrow ac$
$-9 + 1 = -8 \rightarrow b$

$\frac{-9}{3} = \frac{-3}{1}$ ***and*** $\frac{1}{3}$

5 e $30 - 54a - 12a^2$

$= -12a^2 - 54a + 30$

$= -6(2a^2 + 9a - 5)$

$= -6(2a - 1)(a + 5)$

$ac = -10, b = 9$

$-1 \times 10 = -10 \rightarrow ac$
$-1 + 10 = 9 \rightarrow b$

$\frac{-1}{2}$ ***and*** $\frac{10}{2} = \frac{5}{1}$

f $32 + 16b - 6b^2$

$= -6b^2 + 16b + 32$

$= -2(3b^2 - 8b - 16)$

$= -2(b - 4)(3b + 4)$

$ac = -48, b = -8$

$-12 \times 4 = -48 \rightarrow ac$
$-12 + 4 = -8 \rightarrow b$

$\frac{-12}{3} = \frac{-4}{1}$ ***and*** $\frac{4}{3}$

g $63x + 18x^2 - 36$

$= 18x^2 + 63x - 36$

$= 9(2x^2 + 7x - 4)$

$= 9(2x - 1)(x + 4)$

$ac = -8, b = 7$

$-1 \times 8 = -8 \rightarrow ac$
$-1 + 8 = 7 \rightarrow b$

$\frac{-1}{2}$ ***and*** $\frac{8}{2} = \frac{4}{1}$

h $18 - 27b^2 - 45b$

$= -27b^2 - 45b + 18$

$= -9(3b^2 + 5b - 2)$

$= -9(3b - 1)(b + 2)$

$ac = -6, b = 5$

$-1 \times 6 = -6 \rightarrow ac$
$-1 + 6 = 5 \rightarrow b$

$\frac{-1}{3}$ ***and*** $\frac{6}{3} = \frac{2}{1}$

i $8 - 8a - 96a^2$

$= -96a^2 - 8a + 8$

$= -8(12a^2 + a - 1)$

$= -8(4a - 1)(3a + 1)$

$ac = -12, b = 1$

$-3 \times 4 = -12 \rightarrow ac$
$-3 + 4 = 1 \rightarrow b$

$\frac{-3}{12} = \frac{-1}{4}$ ***and*** $\frac{4}{12} = \frac{1}{3}$

Challenge Yourself

1 $2x^2 - 11x + 5 = (2x - 1)(x + k)$

$-1 \times k = 5$

$\therefore k = -5$

2 $x^2 + (k - 5)x - k = (x - 2)(x + 3)$

$(x - 2)(x + 3)$

$= x^2 + 3x - 2x - 6$

$= x^2 + x - 6$

$x^2 + (k - 5)x - k = x^2 + x - 6$

$\therefore k = 6$

3 $x^2 + kx + 3 = (x + 3)(x + m)$

$3m = 3$

$\therefore m = 1$

$(x + 3)(x + 1)$

$= x^2 + x + 3x + 3$

$= x^2 + 4x + 3$

$x^2 + kx + 3 = x^2 + 4x + 3$

$\therefore k = 4$

4 $3x^2 - kx + 12 = (3x + 4)(x + p)$

$4p = 12$

$p = 3$

$(3x + 4)(x + 3)$

$= 3x^2 + 9x + 4x + 12$

$= x^2 + 13x + 12$

$3x^2 - kx + 12 = x^2 + 13x + 12$

$\therefore k = -13$

5 $3x^2 + kx + 5$

$ac = 15, b = k$

$3 \times 5 = 15 \rightarrow ac$
$3 + 5 = 8 \rightarrow b$

$1 \times 15 = 15 \rightarrow ac$
$1 + 15 = 16 \rightarrow b$

$\therefore k = 8 \text{ or } 16$

Completing The Square

Multiple Choice

1 **Option B**

Multiply by coefficient of x^2 is not part of the steps for completing the square.

2 **Option D**

$x^2 + 6x - 5$

$\rightarrow \left(\frac{6}{2}\right)^2 = (3)^2 = 9$

3 **Option A**

$a^2 + 8a - 5$

$\rightarrow \left(\frac{8}{2}\right)^2 = (4)^2$

$= a^2 + 8a + (4)^2 - (4)^2 - 5$

$= [a^2 + 8a + (4)^2] - 16 - 5$

$= (a + 4)^2 - 21$

4 **Option C**

$2p^2 + 12p + 7$

When completing the square, the coefficient of the squared term must be 1. Factoring out the 2 will be the first step.

5 **Option D**

$2x^2 + 4x$

$= 2(x^2 + 2x)$

$\rightarrow \left(\frac{2}{1}\right)^2 = (1)^2 = 1$

$= 2(x^2 + 2x)$

$= 2(x^2 + 2x + (1)^2 - (1)^2)$

$= 2((x + 2)^2 - 1)$

$= 2(x + 2)^2 -2$

Time To Practice

1 **a** $x^2 + 12x$

$\rightarrow \left(\frac{12}{2}\right)^2 = (6)^2$

$= x^2 + 12x + (6)^2 - (6)^2$

$= [x^2 + 12x + (6)^2] - 36$

$= (x + 6)^2 - 36$

$= (x + 6 - 6)(x + 6 + 6)$

$= x(x + 12)$

b $a^2 - 10a$

$\rightarrow \left(\frac{10}{2}\right)^2 = (5)^2$

$= a^2 - 10a + (5)^2 - (5)^2$

$= [a^2 - 10a + (5)^2] - 25$

$= (a - 5)^2 - 25$

$= (a - 5 - 5)(a - 5 + 5)$

$= a(a - 10)$

c $b^2 - 2b$

$\rightarrow \left(\frac{2}{2}\right)^2 = (1)^2$

$= b^2 - 2b + (1)^2 - (1)^2$

$= [b^2 - 2b + (1)^2] - 1$

$= (b - 1)^2 - 1$

$= (b - 1 - 1)(b - 1 + 1)$

$= b(b - 2)$

d $x^2 + 8x$

$\rightarrow \left(\frac{8}{2}\right)^2 = (4)^2$

$= x^2 + 8x + (4)^2 - (4)^2$

$= [x^2 + 8x + (4)^2] - 16$

$= (x + 4)^2 - 16$

$= (x + 4 - 4)(x + 4 + 4)$

$= x(x + 8)$

e $a^2 - 4a$

$\rightarrow \left(\frac{4}{2}\right)^2 = (2)^2$

$= a^2 - 4a + (2)^2 - (2)^2$

$= [a^2 - 4a + (2)^2] - 4$

$= (a - 2)^2 - 4$

$= (a - 2 - 2)(a - 2 + 2)$

$= a(a - 4)$

f $b^2 - 16b$

$\rightarrow \left(\frac{16}{2}\right)^2 = (8)^2$

$= b^2 - 16b + (8)^2 - (8)^2$

$= [b^2 - 16b + (8)^2] - 64$

$= (b - 8)^2 - 64$

$= (b - 8 - 8)(b - 8 + 8)$

$= b(b - 16)$

1 **g** $x^2 + 9x$

$\rightarrow \left(\frac{9}{2}\right)^2$

$= x^2 + 9x + \left(\frac{9}{2}\right)^2 - \left(\frac{9}{2}\right)^2$

$= \left[x^2 + 9x + \left(\frac{9}{2}\right)^2\right] - \left(\frac{9}{2}\right)^2$

$= \left(x + \frac{9}{2}\right)^2 - \left(\frac{9}{2}\right)^2$

$= \left(x + \frac{9}{2} - \frac{9}{2}\right)\left(x + \frac{9}{2} + \frac{9}{2}\right)$

$= x\left(x + \frac{18}{2}\right)$

$= x(x + 9)$

h $a^2 - 7a$

$\rightarrow \left(\frac{7}{2}\right)^2$

$= a^2 - 7a + \left(\frac{7}{2}\right)^2 - \left(\frac{7}{2}\right)^2$

$= \left[a^2 - 7a + \left(\frac{9}{2}\right)^2\right] - \left(\frac{9}{2}\right)^2$

$= \left(a - \frac{7}{2}\right)^2 - \left(\frac{7}{2}\right)^2$

$= \left(a - \frac{7}{2} - \frac{7}{2}\right)\left(a - \frac{7}{2} + \frac{7}{2}\right)$

$= \left(a - \frac{14}{2}\right)a$

$= a(a - 7)$

i $b^2 + 11b$

$\rightarrow \left(\frac{11}{2}\right)^2$

$= b^2 + 11b + \left(\frac{11}{2}\right)^2 - \left(\frac{11}{2}\right)^2$

$= \left[b^2 + 11b + \left(\frac{11}{2}\right)^2\right] - \left(\frac{11}{2}\right)^2$

$= \left(b + \frac{11}{2}\right)^2 - \left(\frac{11}{2}\right)^2$

$= \left(b + \frac{11}{2} - \frac{11}{2}\right)\left(b + \frac{11}{2} + \frac{11}{2}\right)$

$= b\left(b + \frac{22}{2}\right)$

$= b(b + 11)$

j $x^2 - x$

$\rightarrow \left(\frac{1}{2}\right)^2$

$= x^2 - x + \left(\frac{1}{2}\right)^2 - \left(\frac{1}{2}\right)^2$

$= \left[x^2 - x + \left(\frac{1}{2}\right)^2\right] - \left(\frac{1}{2}\right)^2$

$= \left(x - \frac{1}{2}\right)^2 - \left(\frac{1}{2}\right)^2$

$= \left(x - \frac{1}{2} - \frac{1}{2}\right)\left(x - \frac{1}{2} + \frac{1}{2}\right)$

$= \left(x - \frac{2}{2}\right)x$

$= x(x - 1)$

1 **k** $a^2 - 3a$

$$\rightarrow \left(\frac{3}{2}\right)^2$$

$$= a^2 - 3a + \left(\frac{3}{2}\right)^2 - \left(\frac{3}{2}\right)^2$$

$$= \left[a^2 - 3a + \left(\frac{3}{2}\right)^2\right] - \left(\frac{3}{2}\right)^2$$

$$= \left(a - \frac{3}{2}\right)^2 - \left(\frac{3}{2}\right)^2$$

$$= \left(a - \frac{3}{2} - \frac{3}{2}\right)\left(a - \frac{3}{2} + \frac{3}{2}\right)$$

$$= \left(a - \frac{6}{2}\right)a$$

$$= a(a - 3)$$

l $b^2 - 5b$

$$\rightarrow \left(\frac{5}{2}\right)^2$$

$$= b^2 - 5b + \left(\frac{5}{2}\right)^2 - \left(\frac{5}{2}\right)^2$$

$$= \left[b^2 - 5b + \left(\frac{5}{2}\right)^2\right] - \left(\frac{5}{2}\right)^2$$

$$= \left(b - \frac{5}{2}\right)^2 - \left(\frac{5}{2}\right)^2$$

$$= \left(b - \frac{5}{2} - \frac{5}{2}\right)\left(b - \frac{5}{2} + \frac{5}{2}\right)$$

$$= \left(b - \frac{10}{2}\right)b$$

$$= b(b - 5)$$

2 **a** $x^2 + 4x + 1$

$$\rightarrow \left(\frac{4}{2}\right)^2 = (2)^2$$

$$= x^2 + 4x + (2)^2 - (2)^2 + 1$$

$$= [x^2 + 9x + (2)^2] - 3$$

$$= (x + 2)^2 - 3$$

$$= \left(x + 2 - \sqrt{3}\right)\left(x + 2 + \sqrt{3}\right)$$

b $a^2 + 6a + 2$

$$\rightarrow \left(\frac{6}{2}\right)^2 = (3)^2$$

$$= a^2 + 6a + (3)^2 - (3)^2 + 2$$

$$= [a^2 + 6a + (3)^2] - 7$$

$$= (a + 3)^2 - 7$$

$$= \left(a + 3 - \sqrt{7}\right)\left(a + 3 + \sqrt{7}\right)$$

c $b^2 + 2b - 4$

$$\rightarrow \left(\frac{2}{2}\right)^2 = (1)^2$$

$$= b^2 + 2b + (1)^2 - (1)^2 - 4$$

$$= [b^2 + 2b + (1)^2] - 5$$

$$= (b + 1)^2 - 5$$

$$= \left(b + 1 - \sqrt{5}\right)\left(b + 1 + \sqrt{5}\right)$$

d $x^2 + 10x - 4$

$$\rightarrow \left(\frac{10}{2}\right)^2 = (5)^2$$

$$= x^2 + 10x + (5)^2 - (5)^2 - 4$$

$$= [x^2 + 10x + (5)^2] - 29$$

$$= (x + 5)^2 - 29$$

$$= \left(x + 5 - \sqrt{29}\right)\left(x + 5 + \sqrt{29}\right)$$

2 **e** $a^2 - 8a + 13$

$$\rightarrow \left(\frac{8}{2}\right)^2 = (4)^2$$

$$= a^2 - 8a + (4)^2 - (4)^2 + 13$$
$$= [a^2 - 8a + (4)^2] - 3$$
$$= (a - 4)^2 - 3$$
$$= \left(a - 4 - \sqrt{3}\right)\left(a - 4 + \sqrt{3}\right)$$

f $b^2 - 12b + 10$

$$\rightarrow \left(\frac{12}{2}\right)^2 = (6)^2$$

$$= b^2 - 12b + (6)^2 - (6)^2 + 10$$
$$= [b^2 - 12b + (6)^2] - 26$$
$$= (b - 6)^2 - 26$$
$$= \left(b - 6 - \sqrt{26}\right)\left(b - 6 + \sqrt{26}\right)$$

g $x^2 - 4x - 3$

$$\rightarrow \left(\frac{4}{2}\right)^2 = (2)^2$$

$$= x^2 - 4x + (2)^2 - (2)^2 - 3$$
$$= [x^2 - 4x + (2)^2] - 7$$
$$= (x - 2)^2 - 7$$
$$= \left(x - 2 - \sqrt{7}\right)\left(x - 2 + \sqrt{7}\right)$$

h $a^2 - 8a - 5$

$$\rightarrow \left(\frac{8}{2}\right)^2 = (4)^2$$

$$= a^2 - 8a + (4)^2 - (4)^2 - 5$$
$$= [a^2 - 8a + (4)^2] - 21$$
$$= (a - 4)^2 - 21$$
$$= \left(a - 4 - \sqrt{21}\right)\left(a - 4 + \sqrt{21}\right)$$

i $b^2 + 14b + 6$

$$\rightarrow \left(\frac{14}{2}\right)^2 = (7)^2$$

$$= b^2 + 14b + (7)^2 - (7)^2 + 6$$
$$= [b^2 + 14b + (7)^2] - 43$$
$$= (b + 7)^2 - 43$$
$$= \left(b + 7 - \sqrt{43}\right)\left(b + 7 + \sqrt{43}\right)$$

j $7 + 4x - x^2$

$$= -x^2 + 4x + 7$$
$$= -(x^2 - 4x - 7)$$

$$\rightarrow \left(\frac{4}{2}\right)^2 = (2)^2$$

$$= -\left(x^2 - 4x + (2)^2 - (2)^2 - 7\right)$$
$$= -\left([x^2 - 4x + (2)^2] - 11\right)$$
$$= -\left((x - 2)^2 - 11\right)$$
$$= -\left(x - 2 - \sqrt{11}\right)\left(x - 2 + \sqrt{11}\right)$$

2 **k** $2 - 2a - a^2$

$= -a^2 - 2a + 2$

$= -(a^2 + 2a - 2)$

$\rightarrow \left(\frac{2}{2}\right)^2 = (1)^2$

$= -\left(a^2 + 2a + (1)^2 - (1)^2 - 2\right)$

$= -\left([a^2 + 2a + (1)^2] - 3\right)$

$= -\left((a + 1)^2 - 3\right)$

$= -\left(a + 1 - \sqrt{3}\right)\left(a + 1 + \sqrt{3}\right)$

l $10b - 12 - b^2$

$= -b^2 + 10b - 12$

$= -(b^2 - 10b + 12)$

$\rightarrow \left(\frac{10}{2}\right)^2 = (5)^2$

$= -\left(b^2 - 10b + (5)^2 - (5)^2 + 12\right)$

$= -\left([b^2 - 10b + (5)^2] - 13\right)$

$= -\left((b - 5)^2 - 13\right)$

$= -\left(b - 5 - \sqrt{13}\right)\left(b - 5 + \sqrt{13}\right)$

3 **a** $2x^2 + 4x - 4$

$= 2(x^2 + 2x - 2)$

$\rightarrow \left(\frac{2}{2}\right)^2 = (1)^2$

$= 2\left(x^2 + 2x + (1)^2 - (1)^2 - 2\right)$

$= 2\left([x^2 + 2x + (1)^2] - 3\right)$

$= 2\left((x + 1)^2 - 3\right)$

$= 2\left(x + 1 - \sqrt{3}\right)\left(x + 1 + \sqrt{3}\right)$

b $4a^2 - 8a - 20$

$= 4(a^2 - 2a - 5)$

$\rightarrow \left(\frac{2}{2}\right)^2 = (1)^2$

$= 4\left(a^2 - 2a + (1)^2 - (1)^2 - 5\right)$

$= 4\left([a^2 - 2a + (1)^2] - 6\right)$

$= 4\left((a - 1)^2 - 6\right)$

$= 4\left(a - 1 - \sqrt{6}\right)\left(a + 1 + \sqrt{6}\right)$

3 c $5b^2 + 20b - 5$

$= 5(b^2 + 4b - 1)$

$\rightarrow \left(\frac{4}{2}\right)^2 = (2)^2$

$= 5\left(b^2 + 4b + (2)^2 - (2)^2 - 1\right)$

$= 5\left([b^2 + 4b + (2)^2] - 5\right)$

$= 5\left((b + 2)^2 - 5\right)$

$= 5\left(b + 2 - \sqrt{5}\right)\left(b + 2 + \sqrt{5}\right)$

d $3x^2 - 12x - 39$

$= 3(x^2 - 4x - 13)$

$\rightarrow \left(\frac{4}{2}\right)^2 = (2)^2$

$= 3\left(x^2 - 4x + (2)^2 - (2)^2 - 13\right)$

$= 3\left([x^2 - 4x + (2)^2] - 17\right)$

$= 3\left((x - 2)^2 - 17\right)$

$= 3\left(x - 2 - \sqrt{17}\right)\left(x - 2 + \sqrt{17}\right)$

e $5a^2 - 30a + 10$

$= 5(a^2 - 6a + 2)$

$\rightarrow \left(\frac{6}{2}\right)^2 = (3)^2$

$= 5\left(a^2 - 6a + (3)^2 - (3)^2 + 2\right)$

$= 5\left([a^2 - 6a + (3)^2] - 7\right)$

$= 5\left((a - 3)^2 - 7\right)$

$= 5\left(a - 3 - \sqrt{7}\right)\left(a - 3 + \sqrt{7}\right)$

f $2b^2 - 28b + 94$

$= 2(b^2 - 14b + 47)$

$\rightarrow \left(\frac{14}{2}\right)^2 = (7)^2$

$= 2\left(b^2 - 14b + (7)^2 - (7)^2 + 47\right)$

$= 2\left([b^2 - 14b + (7)^2] - 2\right)$

$= 2\left((b - 7)^2 - 2\right)$

$= 2\left(b - 7 - \sqrt{2}\right)\left(b - 7 + \sqrt{2}\right)$

3 **g** $-3x^2 - 30x - 18$

$= -3(x^2 + 10x + 6)$

$\rightarrow \left(\frac{10}{2}\right)^2 = (5)^2$

$= -3\left(x^2 + 10x + (5)^2 - (5)^2 + 6\right)$

$= -3\left([x^2 + 10x + (5)^2] - 19\right)$

$= -3\left((x + 5)^2 - 19\right)$

$= -3\left(x + 5 - \sqrt{19}\right)\left(x + 5 + \sqrt{19}\right)$

h $14 + 8a - 2a^2 = -2a^2 + 8a + 14$

$= -2(a^2 - 4a - 7)$

$\rightarrow \left(\frac{4}{2}\right)^2 = (2)^2$

$= -2\left(a^2 - 4a + (2)^2 - (2)^2 - 7\right)$

$= -2\left([a^2 - 4a + (2)^2] - 11\right)$

$= -2\left((a - 2)^2 - 11\right)$

$= -2\left(a - 2 - \sqrt{11}\right)\left(a - 2 + \sqrt{11}\right)$

i $8b + 16 - 4b^2 = -4b^2 + 8b + 16$

$= -4(b^2 - 2b - 4)$

$\rightarrow \left(\frac{2}{2}\right)^2 = (1)^2$

$= -4\left(b^2 - 2b + (1)^2 - (1)^2 - 4\right)$

$= -4\left([b^2 - 2b + (1)^2] - 5\right)$

$= -4\left((b - 1)^2 - 5\right)$

$= -4\left(b - 1 - \sqrt{5}\right)\left(b - 1 + \sqrt{5}\right)$

4 a

$$3x^2 - 4x - 1 = 3\left(x^2 - \frac{4}{3}x - \frac{1}{3}\right)$$

$$\rightarrow \left(\frac{4}{6}\right)^2 = \left(\frac{2}{3}\right)^2$$

$$= 3\left(x^2 - \frac{4}{3}x + \left(\frac{2}{3}\right)^2 - \left(\frac{2}{3}\right)^2 - \frac{1}{3}\right)$$

$$= 3\left(\left[x^2 - \frac{4}{3}x + \left(\frac{2}{3}\right)^2\right] - \frac{4}{9} - \frac{3}{9}\right)$$

$$= 3\left(\left(x - \frac{2}{3}\right)^2 - \frac{7}{9}\right)$$

$$= 3\left(x - \frac{2}{3} - \frac{\sqrt{7}}{3}\right)\left(x - \frac{2}{3} + \frac{\sqrt{7}}{3}\right)$$

$$= 3\left(x - \frac{2 - \sqrt{7}}{3}\right)\left(x - \frac{2 + \sqrt{7}}{3}\right)$$

4 **b** $5a^2 - 2a - 1 = 5\left(a^2 - \frac{2}{5}a - \frac{1}{5}\right)$

$\rightarrow \left(\frac{2}{10}\right)^2 = \left(\frac{1}{5}\right)^2$

$= 5\left(a^2 - \frac{2}{5}a + \left(\frac{1}{5}\right)^2 - \left(\frac{1}{5}\right)^2 - \frac{1}{5}\right)$

$= 5\left(\left[a^2 - \frac{2}{5}a + \left(\frac{1}{5}\right)^2\right] - \frac{1}{25} - \frac{5}{25}\right)$

$= 5\left(\left(a - \frac{1}{5}\right)^2 - \frac{6}{25}\right)$

$= 5\left(a - \frac{1}{5} - \frac{\sqrt{6}}{5}\right)\left(a - \frac{1}{5} + \frac{\sqrt{6}}{5}\right)$

$= 5\left(a - \frac{1 - \sqrt{6}}{5}\right)\left(a - \frac{1 + \sqrt{6}}{5}\right)$

4 **c**

$$8b^2 - 12b - 1 = 8\left(b^2 - \frac{12}{8}b - \frac{1}{8}\right)$$

$$= 8\left(b^2 - \frac{3}{2}b - \frac{1}{8}\right)$$

$$\rightarrow \left(\frac{3}{4}\right)^2$$

$$= 8\left(b^2 - \frac{3}{2}b + \left(\frac{3}{4}\right)^2 - \left(\frac{3}{4}\right)^2 - \frac{1}{8}\right)$$

$$= 8\left(\left[b^2 - \frac{3}{2}b + \left(\frac{3}{4}\right)^2\right] - \frac{9}{16} - \frac{2}{16}\right)$$

$$= 8\left(\left(b - \frac{3}{4}\right)^2 - \frac{11}{16}\right)$$

$$= 8\left(b - \frac{3}{4} - \frac{\sqrt{11}}{4}\right)\left(b - \frac{3}{4} + \frac{\sqrt{11}}{4}\right)$$

$$= 8\left(b - \frac{3 - \sqrt{11}}{4}\right)\left(b - \frac{3 + \sqrt{11}}{4}\right)$$

4 d

$$-2x^2 - 10x - 11 = -2\left(x^2 + 5x + \frac{11}{2}\right)$$

$$\rightarrow \left(\frac{5}{2}\right)^2$$

$$= -2\left(x^2 + 5x + \left(\frac{5}{2}\right)^2 - \left(\frac{5}{2}\right)^2 + \frac{11}{2}\right)$$

$$= -2\left(\left[x^2 + 5x + \left(\frac{5}{2}\right)^2\right] - \frac{25}{4} + \frac{22}{4}\right)$$

$$= -2\left(\left(x + \frac{5}{2}\right)^2 - \frac{3}{4}\right)$$

$$= -2\left(x + \frac{5}{2} - \frac{\sqrt{3}}{2}\right)\left(x + \frac{5}{2} + \frac{\sqrt{3}}{2}\right)$$

$$= -2\left(x + \frac{5 - \sqrt{3}}{2}\right)\left(x + \frac{5 + \sqrt{3}}{2}\right)$$

4 **e** $$3a^2 - 2a - 2 = 3\left(a^2 - \frac{2}{3}a - \frac{2}{3}\right)$$

$$\rightarrow \left(\frac{2}{6}\right)^2 = \left(\frac{1}{3}\right)^2$$

$$= 3\left(a^2 - \frac{2}{3}a + \left(\frac{1}{3}\right)^2 - \left(\frac{1}{3}\right)^2 - \frac{2}{3}\right)$$

$$= 3\left(\left[a^2 - \frac{2}{3}a + \left(\frac{1}{3}\right)^2\right] - \frac{1}{9} - \frac{6}{9}\right)$$

$$= 3\left(\left(a - \frac{1}{3}\right)^2 - \frac{7}{9}\right)$$

$$= 3\left(a - \frac{1}{3} - \frac{\sqrt{7}}{3}\right)\left(a - \frac{1}{3} + \frac{\sqrt{7}}{3}\right)$$

$$= 3\left(a - \frac{1 - \sqrt{7}}{3}\right)\left(a - \frac{1 + \sqrt{7}}{3}\right)$$

4 f $2b + 2 - 3b^2 = -3b^2 + 2b + 2$

$$= -3\left(b^2 - \frac{2}{3}b - \frac{2}{3}\right)$$

$$\rightarrow \left(\frac{2}{6}\right)^2 = \left(\frac{1}{3}\right)^2$$

$$= -3\left(b^2 - \frac{2}{3}b + \left(\frac{1}{3}\right)^2 - \left(\frac{1}{3}\right)^2 - \frac{2}{3}\right)$$

$$= -3\left(\left[b^2 - \frac{2}{3}b + \left(\frac{1}{3}\right)^2\right] - \frac{1}{9} - \frac{6}{9}\right)$$

$$= -3\left(\left(b - \frac{1}{3}\right)^2 - \frac{7}{9}\right)$$

$$= -3\left(b - \frac{1}{3} - \frac{\sqrt{7}}{3}\right)\left(b - \frac{1}{3} + \frac{\sqrt{7}}{3}\right)$$

$$= -3\left(b - \frac{1 - \sqrt{7}}{3}\right)\left(b - \frac{1 + \sqrt{7}}{3}\right)$$

4 **g** $3 - 2x - 3x^2 = -3x^2 - 2x + 3$

$$= -3\left(x^2 + \frac{2}{3}x - 1\right)$$

$$\rightarrow \left(\frac{2}{6}\right)^2 = \left(\frac{1}{3}\right)^2$$

$$= -3\left(x^2 + \frac{2}{3}x + \left(\frac{1}{3}\right)^2 - \left(\frac{1}{3}\right)^2 - 1\right)$$

$$= -3\left(\left[x^2 + \frac{2}{3}x + \left(\frac{1}{3}\right)^2\right] - \frac{1}{9} - 1\right)$$

$$= -3\left(\left(x + \frac{1}{3}\right)^2 - \frac{1}{9} - \frac{9}{9}\right)$$

$$= -3\left(\left(x + \frac{1}{3}\right)^2 - \frac{10}{9}\right)$$

$$= -3\left(x + \frac{1}{3} - \frac{\sqrt{10}}{3}\right)\left(x + \frac{1}{3} - \frac{\sqrt{10}}{3}\right)$$

$$= -3\left(x + \frac{1 - \sqrt{10}}{3}\right)\left(x + \frac{1 + \sqrt{10}}{3}\right)$$

4 **h** $18a - 9a^2 - 4 = -9a^2 + 18a - 4$

$$= -9\left(a^2 - 2a + \frac{4}{9}\right)$$

$$\rightarrow \left(\frac{2}{2}\right)^2 = (1)^2$$

$$= -9\left(a^2 - 2a + (1)^2 - (1)^2 + \frac{4}{9}\right)$$

$$= -9\left([a^2 - 2a + (1)^2] - 1 + \frac{4}{9}\right)$$

$$= -9\left((a - 1)^2 - \frac{9}{9} + \frac{4}{9}\right)$$

$$= -9\left((a - 1)^2 - \frac{5}{9}\right)$$

$$= -9\left(a - 1 - \frac{\sqrt{5}}{3}\right)\left(a - 1 + \frac{\sqrt{5}}{3}\right)$$

$$= -9\left(a - \frac{3}{3} - \frac{\sqrt{5}}{3}\right)\left(a - \frac{3}{3} + \frac{\sqrt{5}}{3}\right)$$

$$= -9\left(a - \frac{3 - \sqrt{5}}{3}\right)\left(a - \frac{3 + \sqrt{5}}{3}\right)$$

4 i

$$4b^2 + 16b + 3 = 4\left(b^2 + 4b + \frac{3}{4}\right)$$

$$\rightarrow \left(\frac{4}{2}\right)^2 = (2)^2$$

$$= 4\left(b^2 + 4b + (2)^2 - (2)^2 + \frac{3}{4}\right)$$

$$= 4\left([b^2 + 4b + (2)^2] - 4 + \frac{3}{4}\right)$$

$$= 4\left((b + 2)^2 - \frac{16}{4} + \frac{3}{4}\right)$$

$$= 4\left((b + 2)^2 - \frac{13}{4}\right)$$

$$= 4\left(b + 2 - \frac{\sqrt{13}}{2}\right)\left(b + 2 + \frac{\sqrt{13}}{2}\right)$$

$$= 4\left(b + \frac{4}{2} - \frac{\sqrt{13}}{2}\right)\left(b + \frac{4}{2} + \frac{\sqrt{13}}{2}\right)$$

$$= 4\left(b + \frac{4 - \sqrt{13}}{2}\right)\left(b + \frac{4 + \sqrt{13}}{2}\right)$$

Challenge Yourself

1 **a** $x^2 - 4x + \frac{3}{2}$

$\rightarrow \left(\frac{4}{2}\right)^2 = (2)^2$

$= x^2 - 4x + (2)^2 - (2)^2 + \frac{3}{2}$

$= [x^2 - 4x + (2)^2] - 4 + \frac{3}{2}$

$= (x-2)^2 - \frac{8}{2} + \frac{3}{2}$

$= (x-2)^2 - \frac{5}{2}$

$= \left(x - 2 - \frac{\sqrt{5}}{\sqrt{2}}\right)\left(x - 2 + \frac{\sqrt{5}}{\sqrt{2}}\right)$

$\rightarrow \frac{\sqrt{5}}{\sqrt{2}} = \frac{\sqrt{5}}{\sqrt{2}} \times \frac{\sqrt{2}}{\sqrt{2}} = \frac{\sqrt{10}}{2}$

$= \left(x - 2 - \frac{\sqrt{10}}{2}\right)\left(x - 2 + \frac{\sqrt{10}}{2}\right)$

$= \left(x - \frac{4}{2} - \frac{\sqrt{10}}{2}\right)\left(x - \frac{4}{2} + \frac{\sqrt{10}}{2}\right)$

$= \left(x - \frac{4 - \sqrt{10}}{2}\right)\left(x - \frac{4 + \sqrt{10}}{2}\right)$

1 b

$$-x^2 + 6x - \frac{16}{3} = -\left(x^2 - 6x + \frac{16}{3}\right)$$

$$\rightarrow \left(\frac{6}{2}\right)^2 = (3)^2$$

$$= -\left(x^2 - 6x + (3)^2 - (3)^2 + \frac{16}{3}\right)$$

$$= -\left([x^2 - 6x + (3)^2] - 9 + \frac{16}{3}\right)$$

$$= -\left((x-3)^2 - \frac{27}{3} + \frac{16}{3}\right)$$

$$= -\left((x-3)^2 - \frac{11}{3}\right)$$

$$= -\left(x - 3 - \frac{\sqrt{11}}{\sqrt{3}}\right)\left(x - 3 - \frac{\sqrt{11}}{\sqrt{33}}\right)$$

$$\rightarrow \frac{\sqrt{11}}{\sqrt{3}} = \frac{\sqrt{11}}{\sqrt{3}} \times \frac{\sqrt{3}}{\sqrt{3}} = \frac{\sqrt{33}}{3}$$

$$= -\left(x - 3 - \frac{\sqrt{33}}{3}\right)\left(x - 3 + \frac{\sqrt{33}}{3}\right)$$

$$= -\left(x - \frac{9}{3} - \frac{\sqrt{33}}{3}\right)\left(x - \frac{9}{3} + \frac{\sqrt{33}}{3}\right)$$

$$= -\left(x - \frac{9 - \sqrt{33}}{3}\right)\left(x - \frac{9 + \sqrt{33}}{3}\right)$$

1 c $$-3x^2 - 4x + \frac{13}{6} = -3\left(x^2 + \frac{4}{3}x - \frac{13}{18}\right)$$

$$\rightarrow \left(\frac{4}{6}\right)^2 = \left(\frac{2}{3}\right)^2$$

$$= -3\left(x^2 + \frac{4}{3}x + \left(\frac{2}{3}\right)^2 - \left(\frac{2}{3}\right)^2 - \frac{13}{18}\right)$$

$$= -3\left(\left[x^2 + \frac{4}{3}x + \left(\frac{2}{3}\right)^2\right] - \frac{4}{9} - \frac{13}{18}\right)$$

$$= -3\left(\left(x + \frac{2}{3}\right)^2 - \frac{8}{18} - \frac{13}{18}\right)$$

$$= -3\left(\left(x + \frac{2}{3}\right)^2 - \frac{21}{18}\right) = -3\left(\left(x + \frac{2}{3}\right)^2 - \frac{7}{6}\right)$$

$$= -3\left(x + \frac{2}{3} - \frac{\sqrt{7}}{\sqrt{6}}\right)\left(x + \frac{2}{3} + \frac{\sqrt{7}}{\sqrt{6}}\right)$$

$$\rightarrow \frac{\sqrt{7}}{\sqrt{6}} = \frac{\sqrt{7}}{\sqrt{6}} \times \frac{\sqrt{6}}{\sqrt{6}} = \frac{\sqrt{42}}{6}$$

$$= -3\left(x + \frac{2}{3} - \frac{\sqrt{42}}{6}\right)\left(x + \frac{2}{3} + \frac{\sqrt{42}}{6}\right)$$

$$= -3\left(x + \frac{4}{6} - \frac{\sqrt{42}}{6}\right)\left(x + \frac{4}{6} + \frac{\sqrt{42}}{6}\right)$$

$$= -3\left(x + \frac{4 - \sqrt{42}}{6}\right)\left(x + \frac{4 + \sqrt{42}}{6}\right)$$

1 d $2a^2 + 2a - \frac{25}{6} = 2\left(a^2 + a - \frac{25}{12}\right)$

$\rightarrow \left(\frac{1}{2}\right)^2$

$= 2\left(a^2 + a + \left(\frac{1}{2}\right)^2 - \left(\frac{1}{2}\right)^2 - \frac{25}{12}\right)$

$= 2\left(\left[a^2 + a + \left(\frac{1}{2}\right)^2\right] - \frac{1}{4} - \frac{25}{12}\right)$

$= 2\left(\left(a + \frac{1}{2}\right)^2 - \frac{3}{12} - \frac{25}{12}\right)$

$= 2\left(\left(a + \frac{1}{2}\right)^2 - \frac{28}{12}\right) = 2\left(\left(a + \frac{1}{2}\right)^2 - \frac{7}{3}\right)$

$= 2\left(a + \frac{1}{2} - \frac{\sqrt{7}}{\sqrt{3}}\right)\left(a + \frac{1}{2} + \frac{\sqrt{7}}{\sqrt{3}}\right)$

$\rightarrow \frac{\sqrt{7}}{\sqrt{3}} = \frac{\sqrt{7}}{\sqrt{3}} \times \frac{\sqrt{3}}{\sqrt{3}} = \frac{\sqrt{21}}{3}$

$= 2\left(a + \frac{1}{2} - \frac{\sqrt{21}}{3}\right)\left(a + \frac{1}{2} + \frac{\sqrt{21}}{3}\right)$

$= 2\left(a + \frac{3 - 2\sqrt{21}}{6}\right)\left(a + \frac{3 + 2\sqrt{21}}{6}\right)$

1 e $$3a^2 - 2a - \frac{4}{15} = 3\left(a^2 - \frac{2}{3}a - \frac{4}{45}\right)$$

$$\rightarrow \left(\frac{2}{6}\right)^2 = \left(\frac{1}{3}\right)^2$$

$$= 3\left(a^2 - \frac{2}{3}a + \left(\frac{1}{3}\right)^2 - \left(\frac{1}{3}\right)^2 - \frac{4}{45}\right)$$

$$= 3\left(\left[a^2 - \frac{2}{3}a + \left(\frac{1}{3}\right)^2\right] - \frac{1}{9} - \frac{4}{45}\right)$$

$$= 3\left(\left(a - \frac{1}{3}\right)^2 - \frac{5}{45} - \frac{9}{45}\right)$$

$$= 3\left(\left(a - \frac{1}{3}\right)^2 - \frac{9}{45}\right) = \left(\left(a - \frac{1}{3}\right)^2 - \frac{1}{5}\right)$$

$$= 3\left(a - \frac{1}{3} - \frac{1}{\sqrt{5}}\right)\left(a - \frac{1}{3} + \frac{1}{\sqrt{5}}\right)$$

$$\rightarrow \frac{1}{\sqrt{5}} = \frac{1}{\sqrt{5}} \times \frac{\sqrt{5}}{\sqrt{5}} = \frac{\sqrt{5}}{5}$$

$$= 3\left(a - \frac{1}{3} - \frac{\sqrt{5}}{5}\right)\left(a - \frac{1}{3} + \frac{\sqrt{5}}{5}\right)$$

$$= 3\left(a - \frac{5 - 3\sqrt{5}}{15}\right)\left(a - \frac{5 + 3\sqrt{5}}{15}\right)$$

1 f $$-6x^2+2x+\frac{1}{3}=-6\left(x^2-\frac{2}{6}x-\frac{1}{18}\right)=-6\left(x^2-\frac{1}{3}x-\frac{1}{18}\right)$$

$$\rightarrow\left(\frac{1}{6}\right)^2$$

$$=-6\left(x^2-\frac{1}{3}x+\left(\frac{1}{6}\right)^2-\left(\frac{1}{6}\right)^2-\frac{1}{18}\right)$$

$$=-6\left(\left[x^2-\frac{1}{3}x+\left(\frac{1}{6}\right)^2\right]-\frac{1}{36}-\frac{1}{18}\right)$$

$$=-6\left(\left(x-\frac{1}{6}\right)^2-\frac{1}{36}-\frac{2}{36}\right)$$

$$=-6\left(\left(x-\frac{1}{6}\right)^2-\frac{3}{36}\right)=-6\left(\left(x-\frac{1}{6}\right)^2-\frac{1}{12}\right)$$

$$=-6\left(x-\frac{1}{6}-\frac{1}{\sqrt{12}}\right)\left(x-\frac{1}{6}+\frac{1}{\sqrt{12}}\right)$$

$$\rightarrow\frac{1}{\sqrt{12}}=\frac{1}{2\sqrt{3}}=\frac{1}{2\sqrt{3}}\times\frac{\sqrt{3}}{\sqrt{3}}=\frac{\sqrt{3}}{6}$$

$$=-6\left(x-\frac{1}{6}-\frac{\sqrt{3}}{6}\right)\left(x-\frac{1}{6}+\frac{\sqrt{3}}{6}\right)$$

$$=-6\left(x-\frac{1-\sqrt{3}}{6}\right)\left(x-\frac{1+\sqrt{3}}{6}\right)$$

2 $x^2 + 8x + c$

$$\rightarrow \left(\frac{8}{2}\right)^2 = (4)^2 = 16$$

$x^2 + 8x + 16$ is a perfect square trinomial.

$\therefore c = 16$

3 $-5x^2 - 20x - c = -5(x+2)^2$

$$-5(x+2)^2$$
$$= -5(x^2 + 4x + 4)$$
$$= -5x^2 - 20x - 20$$

$\therefore c = 20$

4 $3x^2 - 24x + 57 = 3(x-b)^2 + c$

$$3x^2 - 24x + 57$$
$$= 3(x^2 - 8x + 19)$$

$$\rightarrow \left(\frac{8}{2}\right)^2 = (4)^2$$

$$= 3(x^2 - 8x + (4)^2 - (4)^2 + 19)$$
$$= 3([x^2 - 8x + (4)^2] + 3)$$
$$= 3((x-4)^2 + 3)$$
$$= 3(x-4)^2 + 9$$

$\therefore b = 4 \text{ and } c = 9$

5 $2x^2 + kx + 4 = 2\left(x + \frac{5}{2}\right)^2 - \frac{17}{2}$

$$2\left(x + \frac{5}{2}\right)^2 - \frac{17}{2}$$
$$= 2\left(x^2 + 5x + \frac{25}{4}\right) - \frac{17}{2}$$
$$= 2x^2 + 10x + \frac{25}{2} - \frac{17}{2}$$
$$= 2x^2 + 10x + \frac{8}{2}$$
$$= 2x^2 + 10x + 4$$

$\therefore k = 10$

Solving
Quadratics

Square Root Method

Multiple Choice

1 **Option B**

$x^2 - 64 = 0$

$x^2 = 64$

$x = \pm\sqrt{64} = \pm 8$

2 **Option C**

$2x^2 - 18 = 0$

$2x^2 = 18$

$x^2 = 9$

$x = \pm\sqrt{9} = \pm 3$

3 **Option B**

$4x^2 - 1 = 63$

$4x^2 = 64$

$x^2 = 16$

$x = \pm 4$

4 **Option D**

$x^2 + 25 = 0$

$x^2 = -25$

$x = \pm\sqrt{-25}$

No Solution

5 **Option A**

$2(x+1)^2 = 32$

$(x+1)^2 = 16$

$x + 1 = \pm\sqrt{16} = \pm 4$

$x + 1 = -4 \quad or \quad x + 1 = 4$

$x = -5 \qquad\qquad x = 3$

Time To Practice

1 a
$x^2 - 16 = 0$
$x^2 = 16$
$x = \pm\sqrt{16}$
$x = \pm 4$

b
$a^2 - 64 = 0$
$a^2 = 64$
$a = \pm\sqrt{64}$
$a = \pm 8$

c
$b^2 - 20 = 5$
$b^2 = 25$
$b = \pm\sqrt{25}$
$a = \pm 5$

d
$x^2 - 121 = 0$
$x^2 = 121$
$x = \pm\sqrt{121}$
$x = \pm 11$

e
$6a^2 - 54 = 0$
$6a^2 = 54$
$a^2 = 9$
$a = \pm\sqrt{9}$
$a = \pm 3$

f
$16b^2 - 16 = 0$
$16b^2 = 16$
$b^2 = 1$
$b = \pm\sqrt{1}$
$b = \pm 1$

g
$200 - 2x^2 = 0$
$200 = 2x^2$
$x^2 = 100$
$x = \pm\sqrt{100}$
$x = \pm 10$

h
$72 - 2a^2 = 0$
$72 = 2a^2$
$a^2 = 36$
$a = \pm\sqrt{36}$
$a = \pm 6$

i
$-7b^2 + 49 = 0$
$49 = 7b^2$
$b^2 = 7$
$b = \pm\sqrt{7}$

j
$x^2 + 9 = 0$
$x^2 = -9$
$x = \pm\sqrt{-9}$
No Solution

1 k $a^2 + 4 = 0$

$a^2 = -4$

$a = \pm\sqrt{-4}$

$No\ Solution$

l $8b^2 + 72 = 0$

$8b^2 = -72$

$b^2 = -9$

$b = \pm\sqrt{-9}$

$No\ Solution$

m $x^2 - 10 = 0$

$x^2 = 10$

$x = \pm\sqrt{10}$

n $3a^2 - 45 = 0$

$3a^2 = 45$

$a^2 = 15$

$a = \pm\sqrt{15}$

o $22 - 2b^2 = 0$

$2b^2 = 22$

$b^2 = 11$

$b = \pm\sqrt{11}$

2 a $x^2 = \frac{1}{4}$

$x = \pm\sqrt{\frac{1}{4}}$

$x = \pm\frac{1}{2}$

b $4a^2 = \frac{9}{16}$

$a^2 = \frac{9}{16} \times \frac{1}{4}$

$a^2 = \frac{9}{64}$

$a = \pm\sqrt{\frac{9}{64}}$

$a = \pm\frac{3}{8}$

2 **c**

$$4b^2 - \frac{1}{9} = 0$$

$$4b^2 = \frac{1}{9}$$

$$b^2 = \frac{1}{9} \times \frac{1}{4}$$

$$b^2 = \frac{1}{36}$$

$$b = \pm\sqrt{\frac{1}{36}}$$

$$b = \pm\frac{1}{6}$$

d

$$9x^2 - 4 = 0$$

$$9x^2 = 4$$

$$x^2 = \frac{4}{9}$$

$$x = \pm\sqrt{\frac{4}{9}}$$

$$x = \pm\frac{2}{3}$$

e

$$36a^2 - 49 = 0$$

$$36a^2 = 49$$

$$a^2 = \frac{49}{36}$$

$$a = \pm\sqrt{\frac{49}{36}}$$

$$a = \pm\frac{7}{6}$$

f

$$100 - 81b^2 = 0$$

$$100 = 81b^2$$

$$b^2 = \frac{100}{81}$$

$$b = \pm\sqrt{\frac{100}{81}}$$

$$b = \pm\frac{10}{9}$$

3 **a**

$$(x + 4)^2 = 0$$

$$x + 4 = 0$$

$$x = -4$$

b

$$(a - 5)^2 = 0$$

$$a - 5 = 0$$

$$a = 5$$

c

$$(2 - b)^2 = 0$$

$$2 - b = 0$$

$$b = 2$$

d

$$(2x + 3)^2 = 0$$

$$2x + 3 = 0$$

$$2x = -3$$

$$x = -\frac{3}{2}$$

3 **e** $(7-6a)^2=0$

$7-6a=0$

$7=6a$

$a=\frac{7}{6}$

f $(-4b-1)^2=0$

$-4b-1=0$

$-4b=1$

$b=-\frac{1}{4}$

g $(x+3)^2-9=0$

$(x+3)^2=9$

$x+3=\pm\sqrt{9}$

$x+3=\pm 3$

$x+3=-3 \quad or \quad x+3=3$

$x=-6 \quad or \quad x=0$

h $(a-5)^2-16=0$

$(a-5)^2=16$

$a-5=\pm\sqrt{16}$

$a-5=\pm 4$

$a-5=-4 \quad or \quad a-5=4$

$a=1 \quad or \quad a=9$

i $(7-b)^2-36=0$

$(7-b)^2=36$

$7-b=\pm\sqrt{36}$

$7-b=\pm 6$

$7-b=-6 \quad or \quad 7-b=6$

$b=13 \quad or \quad b=1$

4 **a** $2(x+1)^2-8=0$

$2(x+1)^2=8$

$(x+1)^2=4$

$x+1=\pm\sqrt{4}$

$x+1=\pm 2$

$x+1=-2 \quad or \quad x+1=2$

$x=-3 \quad or \quad x=1$

b $5(a+3)^2-20=0$

$5(a+3)^2=20$

$(a+3)^2=4$

$a+3=\pm\sqrt{4}$

$a+3=\pm 2$

$a+3=-2 \quad or \quad a+3=2$

$a=-5 \quad or \quad a=-1$

4 **c** $3(b-5)^2-27=0$

$3(b-5)^2=27$

$(b-5)^2=9$

$b-5=\pm\sqrt{9}$

$b-5=\pm3$

$b-5=-3$ *or* $b-5=3$

$a=2$ *or* $b=8$

d $-3(x-4)^2+12=0$

$-3(x-4)^2=-12$

$(x-4)^2=4$

$x-4=\pm\sqrt{4}$

$x-4=\pm2$

$x-4=-2$ *or* $x-4=2$

$x=2$ *or* $x=6$

e $4(2-a)^2-16=0$

$4(2-a)^2=16$

$(2-a)^2=4$

$2-a=\pm\sqrt{4}$

$2-a=\pm2$

$2-a=-2$ *or* $2-a=2$

$a=4$ *or* $a=0$

f $-(3b-4)^2+9=0$

$-(3b-4)^2=-9$

$(3b-4)^2=9$

$3b-4=\pm\sqrt{9}$

$3b-4=\pm3$

$3b-4=-3$ *or* $3b-4=3$

$3b=1$ *or* $3b=7$

$b=\frac{1}{3}$ *or* $b=\frac{7}{3}$

g $25-(2x+1)^2=0$

$-(2x+1)^2=-25$

$(2x+1)^2=25$

$2x+1=\pm\sqrt{25}$

$2x+1=\pm5$

$2x+1=-5$ *or* $2x+1=5$

$2x=-6$ *or* $2x=4$

$x=-3$ *or* $x=2$

h $6(7-2a)^2=24$

$(7-2a)^2=4$

$7-2a=\pm\sqrt{4}$

$7-2a=\pm2$

$7-2a=-2$ *or* $7-2a=2$

$2a=9$ *or* $2a=5$

$a=\frac{9}{2}$ *or* $a=\frac{5}{2}$

4 i
$$-3(4b-5)^2=-75$$
$$(4b-5)^2=25$$
$$4b-5=\pm\sqrt{25}$$
$$4b-5=\pm5$$
$$4b-5=-5 \quad or \quad 4b-5=5$$
$$4b=0 \quad or \quad 4b=10$$
$$b=0 \quad or \quad b=\frac{10}{4}=\frac{5}{2}$$

j
$$2(3x-5)^2-50=0$$
$$2(3x-5)^2=50$$
$$(3x-5)^2=25$$
$$3x-5=\pm\sqrt{25}$$
$$3x-5=\pm5$$
$$3x-5=-5 \quad or \quad 3x-5=5$$
$$3x=0 \quad or \quad 3x=10$$
$$x=0 \quad or \quad x=\frac{10}{3}$$

k
$$-5(1-2a)^2=-80$$
$$(1-2a)^2=16$$
$$1-2a=\pm\sqrt{16}$$
$$1-2a=\pm4$$
$$1-2a=-4 \quad or \quad 1-2a=4$$
$$2a=5 \quad or \quad 2a=-3$$
$$a=\frac{5}{2} \quad or \quad a=-\frac{3}{2}$$

l
$$-6(5b-1)^2=-54$$
$$(5b-1)^2=9$$
$$5b-1=\pm\sqrt{9}$$
$$5b-1=\pm3$$
$$5b-1=-3 \quad or \quad 5b-1=3$$
$$5b=-2 \quad or \quad 5b=4$$
$$b=-\frac{2}{5} \quad or \quad b=\frac{4}{5}$$

Challenge Yourself

1 a
$$-5(1-2x)^2=-40$$
$$(1-2x)^2=8$$
$$1-2x=\pm\sqrt{8}$$
$$1-2x=\pm2\sqrt{2}$$
$$1\pm2\sqrt{2}=2x$$
$$x=\frac{1\pm2\sqrt{2}}{2}$$

b
$$-2(2m-1)^2=-36$$
$$(2m-1)^2=18$$
$$2m-1=\pm\sqrt{18}$$
$$2m-1=\pm3\sqrt{2}$$
$$2m=\pm3\sqrt{2}+1=1\pm3\sqrt{2}$$
$$m=\frac{1\pm3\sqrt{2}}{2}$$

1 c

$$\frac{1}{3}(2x-1)^2-5=0$$

$$\frac{1}{3}(2x-1)^2=5$$

$$(2x-1)^2=15$$

$$2x-1=\pm\sqrt{15}$$

$$2x=1\pm\sqrt{15}$$

$$x=\frac{1\pm\sqrt{15}}{2}$$

d

$$-\frac{3}{4}(1-2n)^2+9=0$$

$$-\frac{3}{4}(1-2n)^2=-9$$

$$(1-2n)^2=-9\times-\frac{4}{3}=12$$

$$1-2n=\pm\sqrt{12}$$

$$1-2n=\pm2\sqrt{3}$$

$$1\pm2\sqrt{3}=2n$$

$$n=\frac{1\pm2\sqrt{3}}{2}$$

e

$$\frac{5}{3}(3p+2)^2-40=0$$

$$\frac{5}{3}(3p+2)^2=40$$

$$(3p+2)^2=40\times\frac{3}{5}=24$$

$$3p+2=\pm\sqrt{24}$$

$$3p+2=\pm2\sqrt{6}$$

$$3p=-2\pm2\sqrt{6}$$

$$p=\frac{-2\pm2\sqrt{6}}{3}$$

f

$$-\frac{1}{4}(3x+4)^2+7=0$$

$$-\frac{1}{4}(3x+4)^2=-7$$

$$(3x+4)^2=28$$

$$3x+4=\pm\sqrt{28}$$

$$3x+4=\pm2\sqrt{7}$$

$$3x=-4\pm2\sqrt{7}$$

$$x=\frac{-4\pm2\sqrt{7}}{3}$$

1 **g**

$$\frac{3}{2}\left(x-\frac{1}{3}\right)^2-27=0$$

$$\frac{3}{2}\left(x-\frac{1}{3}\right)^2=27$$

$$\left(x-\frac{1}{3}\right)^2=27\times\frac{2}{3}=18$$

$$x-\frac{1}{3}=\pm\sqrt{18}$$

$$x-\frac{1}{3}=\pm3\sqrt{2}$$

$$x=\frac{1}{3}\pm3\sqrt{2}=\frac{1}{3}\pm\frac{3\sqrt{2}}{1}$$

$$x=\frac{1\pm9\sqrt{2}}{3}$$

h

$$30-\frac{2}{3}\left(q+\frac{1}{2}\right)^2=0$$

$$-\frac{2}{3}\left(q+\frac{1}{2}\right)^2=-30$$

$$\left(q+\frac{1}{2}\right)^2=-30\times-\frac{3}{2}=45$$

$$q+\frac{1}{2}=\pm\sqrt{45}$$

$$q+\frac{1}{2}=\pm3\sqrt{5}$$

$$q=-\frac{1}{2}\pm3\sqrt{5}=-\frac{1}{2}\pm\frac{3\sqrt{5}}{1}$$

$$q=\frac{-1\pm6\sqrt{5}}{2}$$

2

$$(2x-1)^2-(x-k)(x+k)=3x^2-4x+10$$

$$Left\ Side\ =4x^2-4x+1-\left(x^2-k^2\right)$$

$$=4x^2-4x+1-x^2+k^2$$

$$=3x^2-4x+1+k^2$$

$$3x^2-4x+1+k^2=3x^2-4x+10$$

$$\therefore 1+\ k^2=10$$

$$k^2=9$$

$$k=\pm\sqrt{9}$$

$$k=\pm3$$

Completing The Square

Multiple Choice

1 **Option C**

$x^2 + 8x + 5 = 0$

$x^2 + 8x + (4)^2 - (4)^2 + 5 = 0$

$[x^2 + 8x + (4)^2] - 16 + 5 = 0$

$(x + 4)^2 - 11 = 0$

2 **Option A**

$x^2 + 10x + 21 = 0$

$x^2 + 10x + (5)^2 - (5)^2 + 21 = 0$

$[x^2 + 10x + (5)^2] - 25 + 21 = 0$

$(x + 5)^2 - 4 = 0$

3 **Option B**

$3(x - 1)^2 = 27$

$(x - 1)^2 = 9$

$x - 1 = \pm\sqrt{9}$

$x - 1 = \pm 3$

$x - 1 = -3 \quad or \quad x - 1 = 3$

$x = -2 \quad or \quad x = 4$

4 **Option D**

$x^2 = 4x + 1$

$x^2 - 4x - 1 = 0$

$x^2 - 4x + (2)^2 - (2)^2 - 1 = 0$

$[x^2 - 4x + (2)^2] - 4 - 1 = 0$

$(x - 2)^2 - 5 = 0$

$(x - 2)^2 = 5$

$x - 2 = \pm\sqrt{5}$

$x = 2 \pm \sqrt{5}$

5 **Option A**

$x^2 - 8x - 7 = 0$

$x^2 - 8x + (4)^2 - (4)^2 - 7 = 0$

$[x^2 - 8x + (4)^2] - 16 - 7 = 0$

$(x - 4)^2 - 23 = 0$

$(x - 4)^2 = 23$

$x - 4 = \pm\sqrt{23}$

$x = 4 \pm \sqrt{23}$

Time To Practice

1 a $x^2 + 2x - 2 = 0$

$x^2 + 2x + (1)^2 - (1)^2 - 2 = 0$

$[x^2 + 2x + (1)^2] - 3 = 0$

$(x + 1)^2 - 3 = 0$

$(x + 1)^2 = 3$

$x + 1 = \pm\sqrt{3}$

$x = -1 \pm \sqrt{3}$

b $a^2 - 2a - 5 = 0$

$a^2 - 2a + (1)^2 - (1)^2 - 5 = 0$

$[a^2 - 2a + (1)^2] - 6 = 0$

$(a - 1)^2 - 6 = 0$

$(a - 1)^2 = 6$

$a - 1 = \pm\sqrt{6}$

$a = 1 \pm \sqrt{6}$

c $b^2 + 2b - 1 = 0$

$b^2 + 2b + (1)^2 - (1)^2 - 1 = 0$

$[b^2 + 2b + (1)^2] - 2 = 0$

$(b + 1)^2 - 2 = 0$

$(b + 1)^2 = 2$

$b + 1 = \pm\sqrt{2}$

$b = -1 \pm \sqrt{2}$

d $x^2 - 2x - 4 = 0$

$x^2 - 2x + (1)^2 - (1)^2 - 4 = 0$

$[x^2 - 2x + (1)^2] - 5 = 0$

$(x - 1)^2 - 5 = 0$

$(x - 1)^2 = 5$

$x - 1 = \pm\sqrt{5}$

$x = 1 \pm \sqrt{5}$

e $a^2 + 6a + 7 = 0$

$a^2 + 6a + (3)^2 - (3)^2 + 7 = 0$

$[a^2 + 6a + (3)^2] - 2 = 0$

$(a + 3)^2 - 2 = 0$

$(a + 3)^2 = 2$

$a + 3 = \pm\sqrt{2}$

$a = -3 \pm \sqrt{2}$

f $b^2 - 8b + 9 = 0$

$b^2 - 8b + (4)^2 - (4)^2 + 9 = 0$

$[b^2 - 8b + (4)^2] - 7 = 0$

$(b - 4)^2 - 7 = 0$

$(b - 4)^2 = 7$

$b - 4 = \pm\sqrt{7}$

$b = 4 \pm \sqrt{7}$

1 g

$10 - 6x - x^2 = 0$

$-x^2 - 6x + 10 = 0$

$-(x^2 + 6x - 10) = 0$

$x^2 + 6x - 10 = 0$

$x^2 + 6x + (3)^2 - (3)^2 - 10 = 0$

$[x^2 + 6x + (3)^2] - 19 = 0$

$(x + 3)^2 - 19 = 0$

$(x + 3)^2 = 19$

$x + 3 = \pm\sqrt{19}$

$x = -3 \pm \sqrt{19}$

h

$10a + 4 - a^2 = 0$

$-a^2 + 10a + 4 = 0$

$-(a^2 - 10a - 4) = 0$

$a^2 - 10a - 4 = 0$

$a^2 - 10a + (5)^2 - (5)^2 - 4 = 0$

$[a^2 - 10a + (5)^2] - 29 = 0$

$(a - 5)^2 - 29 = 0$

$(a - 5)^2 = 29$

$a - 5 = \pm\sqrt{29}$

$a = 5 \pm \sqrt{29}$

i

$-1 - 12b - b^2 = 0$

$-b^2 - 12b - 1 = 0$

$-(b^2 + 12b + 1) = 0$

$b^2 + 12b + 1 = 0$

$b^2 + 12b + (6)^2 - (6)^2 + 1 = 0$

$[b^2 + 12b + (6)^2] - 35 = 0$

$(b + 6)^2 - 35 = 0$

$(b + 6)^2 = 35$

$b + 6 = \pm\sqrt{35}$

$b = -6 \pm \sqrt{35}$

2 **a** $x^2 - x - 1 = 0$

$x^2 - x + \left(\frac{1}{2}\right)^2 - \left(\frac{1}{2}\right)^2 - 1 = 0$

$\left[x^2 - x + \left(\frac{1}{2}\right)^2\right] - \frac{1}{4} - \frac{4}{4} = 0$

$\left(x - \frac{1}{2}\right)^2 - \frac{5}{4} = 0$

$\left(x - \frac{1}{2}\right)^2 = \frac{5}{4}$

$x - \frac{1}{2} = \pm\frac{\sqrt{5}}{2}$

$x = \frac{1}{2} \pm \frac{\sqrt{5}}{2}$

$x = \frac{1 \pm \sqrt{5}}{2}$

b $a^2 + 3a - 2 = 0$

$a^2 + 3a + \left(\frac{3}{2}\right)^2 - \left(\frac{3}{2}\right)^2 - 2 = 0$

$\left[a^2 + 3a + \left(\frac{3}{2}\right)^2\right] - \frac{9}{4} - \frac{8}{4} = 0$

$\left(a + \frac{3}{2}\right)^2 - \frac{17}{4} = 0$

$\left(a + \frac{3}{2}\right)^2 = \frac{17}{4}$

$a + \frac{3}{2} = \pm\frac{\sqrt{17}}{2}$

$a = -\frac{3}{2} \pm \frac{\sqrt{17}}{2}$

$a = \frac{-3 \pm \sqrt{17}}{2}$

c $b^2 - 5b - 3 = 0$

$b^2 - 5b + \left(\frac{5}{2}\right)^2 - \left(\frac{5}{2}\right)^2 - 3 = 0$

$\left[b^2 - 5b + \left(\frac{5}{2}\right)^2\right] - \frac{25}{4} - \frac{12}{4} = 0$

$\left(b - \frac{5}{2}\right)^2 - \frac{37}{4} = 0$

$\left(b - \frac{5}{2}\right)^2 = \frac{37}{4}$

$b - \frac{5}{2} = \pm\frac{\sqrt{37}}{2}$

$b = \frac{5}{2} \pm \frac{\sqrt{37}}{2}$

$b = \frac{5 \pm \sqrt{37}}{2}$

d $x^2 - 3x + 1 = 0$

$x^2 - 3x + \left(\frac{3}{2}\right)^2 - \left(\frac{3}{2}\right)^2 + 1 = 0$

$\left[x^2 - 3x + \left(\frac{3}{2}\right)^2\right] - \frac{9}{4} + \frac{4}{4} = 0$

$\left(x - \frac{3}{2}\right)^2 - \frac{5}{4} = 0$

$\left(x - \frac{3}{2}\right)^2 = \frac{5}{4}$

$x - \frac{3}{2} = \pm\frac{\sqrt{5}}{2}$

$x = \frac{3}{2} \pm \frac{\sqrt{5}}{2}$

$x = \frac{3 \pm \sqrt{5}}{2}$

2 **e** $a^2 + 5a + 3 = 0$

$a^2 + 5a + \left(\frac{5}{2}\right)^2 - \left(\frac{5}{2}\right)^2 + 3 = 0$

$\left[a^2 + 5a + \left(\frac{5}{2}\right)^2\right] - \frac{25}{4} + \frac{12}{4} = 0$

$\left(a + \frac{5}{2}\right)^2 - \frac{13}{4} = 0$

$\left(a + \frac{5}{2}\right)^2 = \frac{13}{4}$

$a + \frac{5}{2} = \pm\frac{\sqrt{13}}{2}$

$a = -\frac{5}{2} \pm \frac{\sqrt{13}}{2}$

$a = \frac{-5 \pm \sqrt{13}}{2}$

f $b^2 + 7b - 1 = 0$

$b^2 + 7b + \left(\frac{7}{2}\right)^2 - \left(\frac{7}{2}\right)^2 - 1 = 0$

$\left[b^2 + 7b + \left(\frac{7}{2}\right)^2\right] - \frac{49}{4} - \frac{4}{4} = 0$

$\left(b + \frac{7}{2}\right)^2 - \frac{53}{4} = 0$

$\left(b + \frac{7}{2}\right)^2 = \frac{53}{4}$

$b + \frac{7}{2} = \pm\frac{\sqrt{53}}{2}$

$b = -\frac{7}{2} \pm \frac{\sqrt{53}}{2}$

$b = \frac{-7 \pm \sqrt{53}}{2}$

g $x^2 + 3x - 6 = 0$

$x^2 + 3x + \left(\frac{3}{2}\right)^2 - \left(\frac{3}{2}\right)^2 - 6 = 0$

$\left[x^2 + 3x + \left(\frac{3}{2}\right)^2\right] - \frac{9}{4} - \frac{24}{4} = 0$

$\left(x + \frac{3}{2}\right)^2 - \frac{33}{4} = 0$

$\left(x + \frac{3}{2}\right)^2 = \frac{33}{4}$

$x + \frac{3}{2} = \pm\frac{\sqrt{33}}{2}$

$x = -\frac{3}{2} \pm \frac{\sqrt{5}}{2}$

$x = \frac{-3 \pm \sqrt{33}}{2}$

h $a^2 - 9a + 7 = 0$

$a^2 - 9a + \left(\frac{9}{2}\right)^2 - \left(\frac{9}{2}\right)^2 + 7 = 0$

$\left[a^2 - 9a + \left(\frac{9}{2}\right)^2\right] - \frac{81}{4} + \frac{28}{4} = 0$

$\left(a - \frac{9}{2}\right)^2 - \frac{53}{4} = 0$

$\left(a - \frac{9}{2}\right)^2 = \frac{53}{4}$

$a - \frac{9}{2} = \pm\frac{\sqrt{53}}{2}$

$a = \frac{9}{2} \pm \frac{\sqrt{53}}{2}$

$a = \frac{9 \pm \sqrt{53}}{2}$

2 **i** $-b^2 = 20 - 11b \rightarrow b^2 - 11b + 20 = 0$

$$b^2 - 11b + \left(\frac{11}{2}\right)^2 - \left(\frac{11}{2}\right)^2 + 20 = 0$$

$$\left[b^2 - 11b + \left(\frac{11}{2}\right)^2\right] - \frac{121}{4} + \frac{80}{4} = 0$$

$$\left(b - \frac{11}{2}\right)^2 - \frac{41}{4} = 0$$

$$\left(b - \frac{11}{2}\right)^2 = \frac{41}{4}$$

$$b - \frac{11}{2} = \pm\frac{\sqrt{41}}{2}$$

$$b = \frac{11}{2} \pm \frac{\sqrt{41}}{2}$$

$$b = \frac{11 \pm \sqrt{41}}{2}$$

j $5x - x^2 = -10 \rightarrow x^2 - 5x - 10 = 0$

$$x^2 - 5x + \left(\frac{5}{2}\right)^2 - \left(\frac{5}{2}\right)^2 - 10 = 0$$

$$\left[x^2 - 5x + \left(\frac{5}{2}\right)^2\right] - \frac{25}{4} - \frac{40}{4} = 0$$

$$\left(x - \frac{5}{2}\right)^2 - \frac{65}{4} = 0$$

$$\left(x - \frac{5}{2}\right)^2 = \frac{65}{4}$$

$$x - \frac{5}{2} = \pm\frac{\sqrt{65}}{2}$$

$$x = \frac{5}{2} \pm \frac{\sqrt{65}}{2}$$

$$x = \frac{5 \pm \sqrt{65}}{2}$$

2 **k** $11a - 25 = a^2 \rightarrow a^2 - 11a + 25 = 0$

$$a^2 - 11a + \left(\frac{11}{2}\right)^2 - \left(\frac{11}{2}\right)^2 + 25 = 0$$

$$\left[a^2 - 11a + \left(\frac{11}{2}\right)^2\right] - \frac{121}{4} + \frac{100}{4} = 0$$

$$\left(a + \frac{11}{2}\right)^2 - \frac{21}{4} = 0$$

$$\left(a + \frac{11}{2}\right)^2 = \frac{21}{4}$$

$$a + \frac{11}{2} = \pm\frac{\sqrt{21}}{2}$$

$$a = \frac{11}{2} \pm \frac{\sqrt{21}}{2}$$

$$a = \frac{11 \pm \sqrt{21}}{2}$$

l $7b + 9 = -b^2 \rightarrow b^2 + 7b + 9 = 0$

$$b^2 + 7b + \left(\frac{7}{2}\right)^2 - \left(\frac{7}{2}\right)^2 + 9 = 0$$

$$\left[b^2 + 7b + \left(\frac{7}{2}\right)^2\right] - \frac{49}{4} + \frac{36}{4} = 0$$

$$\left(b + \frac{7}{2}\right)^2 - \frac{13}{4} = 0$$

$$\left(b + \frac{7}{2}\right)^2 = \frac{13}{4}$$

$$b + \frac{7}{2} = \pm\frac{\sqrt{13}}{2}$$

$$b = -\frac{7}{2} \pm \frac{\sqrt{13}}{2}$$

$$b = \frac{-7 \pm \sqrt{13}}{2}$$

3 **a** $2x^2 + 4x - 8 = 0$

$2(x^2 + 2x - 4) = 0$

$x^2 + 2x - 4 = 0$

$x^2 + 2x + (1)^2 - (1)^2 - 4 = 0$

$[x^2 + 2x + (1)^2] - 5 = 0$

$(x + 1)^2 = 5$

$x + 1 = \pm\sqrt{5}$

$x = -1 \pm \sqrt{5}$

b $3a^2 + 12a - 6 = 0$

$3(a^2 + 4a - 2) = 0$

$a^2 + 4a - 2 = 0$

$a^2 + 4a + (2)^2 - (2)^2 - 2 = 0$

$[a^2 + 4a + (2)^2] - 6 = 0$

$(a + 2)^2 - 6 = 0$

$a + 2 = \pm\sqrt{6}$

$a = -2 \pm \sqrt{6}$

c $4b^2 - 8b - 8 = 0$

$4(b^2 - 2b - 4) = 0$

$b^2 - 2b - 2 = 0$

$b^2 - 2b + (1)^2 - (1)^2 - 2 = 0$

$[b^2 - 2b + (1)^2] - 3 = 0$

$(b - 1)^2 = 3$

$b - 1 = \pm\sqrt{3}$

$b = 1 \pm \sqrt{3}$

d $5x^2 - 15x - 15 = 0$

$5(x^2 - 3x - 3) = 0$

$x^2 - 3x - 3 = 0$

$$x^2 - 3x + \left(\frac{3}{2}\right)^2 - \left(\frac{3}{2}\right)^2 - 3 = 0$$

$$\left[x^2 - 3x + \left(\frac{3}{2}\right)^2\right] - \frac{9}{4} - \frac{12}{4} = 0$$

$$\left(x - \frac{3}{2}\right)^2 - \frac{21}{4} = 0$$

$$\left(x - \frac{3}{2}\right)^2 = \frac{21}{4}$$

$$x - \frac{3}{2} = \pm\frac{\sqrt{21}}{2}$$

$$x = \frac{3}{2} \pm \frac{\sqrt{21}}{2}$$

$$x = \frac{3 \pm \sqrt{21}}{2}$$

3 **e**

$$2a^2 - 6a + 2 = 0$$
$$2(a^2 - 3a + 1) = 0$$
$$a^2 - 3a + 1 = 0$$

$$a^2 - 3a + \left(\frac{3}{2}\right)^2 - \left(\frac{3}{2}\right)^2 + 1 = 0$$
$$\left[a^2 - 3a + \left(\frac{3}{2}\right)^2\right] - \frac{9}{4} + \frac{4}{4} = 0$$
$$\left(a - \frac{3}{2}\right)^2 - \frac{5}{4} = 0$$
$$\left(a - \frac{3}{2}\right)^2 = \frac{5}{4}$$
$$a - \frac{3}{2} = \pm\frac{\sqrt{5}}{2}$$
$$a = \frac{3}{2} \pm \frac{\sqrt{5}}{2}$$
$$a = \frac{3 \pm \sqrt{5}}{2}$$

f

$$3b^2 - 9b - 3 = 0$$
$$3(b^2 - 3b - 1) = 0$$
$$b^2 - 3b - 1 = 0$$

$$b^2 - 3b + \left(\frac{3}{2}\right)^2 - \left(\frac{3}{2}\right)^2 - 1 = 0$$
$$\left[b^2 - 3b + \left(\frac{3}{2}\right)^2\right] - \frac{9}{4} - \frac{4}{4} = 0$$
$$\left(b - \frac{3}{2}\right)^2 - \frac{13}{4} = 0$$
$$\left(b - \frac{3}{2}\right)^2 = \frac{13}{4}$$
$$b - \frac{3}{2} = \pm\frac{\sqrt{13}}{2}$$
$$b = \frac{3}{2} \pm \frac{\sqrt{13}}{2}$$
$$b = \frac{3 \pm \sqrt{13}}{2}$$

3 **g**

$$-5x^2 - 15x = -25$$
$$-5x^2 - 15x + 25 = 0$$
$$-5(x^2 + 3x - 5) = 0$$
$$x^2 + 3x - 5 = 0$$

$$x^2 + 3x + \left(\frac{3}{2}\right)^2 - \left(\frac{3}{2}\right)^2 - 5 = 0$$
$$\left[x^2 + 3x + \left(\frac{3}{2}\right)^2\right] - \frac{9}{4} - \frac{20}{4} = 0$$
$$\left(x + \frac{3}{2}\right)^2 - \frac{29}{4} = 0$$
$$\left(x + \frac{3}{2}\right)^2 = \frac{29}{4}$$
$$x + \frac{3}{2} = \pm\frac{\sqrt{29}}{2}$$
$$x = -\frac{3}{2} \pm \frac{\sqrt{29}}{2}$$
$$x = \frac{-3 \pm \sqrt{29}}{2}$$

h

$$21 - 7a = 7a^2$$
$$7a^2 + 7a - 21 = 0$$
$$7(a^2 + a - 3) = 0$$
$$a^2 + a - 3 = 0$$

$$a^2 + a + \left(\frac{1}{2}\right)^2 - \left(\frac{1}{2}\right)^2 - 3 = 0$$
$$\left[a^2 + a + \left(\frac{1}{2}\right)^2\right] - \frac{1}{4} - \frac{12}{4} = 0$$
$$\left(a + \frac{1}{2}\right)^2 - \frac{13}{4} = 0$$
$$\left(a + \frac{1}{2}\right)^2 = \frac{13}{4}$$
$$a + \frac{1}{2} = \pm\frac{\sqrt{13}}{2}$$
$$a = -\frac{1}{2} \pm \frac{\sqrt{13}}{2}$$
$$a = \frac{-1 \pm \sqrt{13}}{2}$$

3 **i** $-2b^2 - 18b = 38$

$2b^2 + 18b + 38 = 0$

$2(b^2 + 9b + 19) = 0$

$b^2 + 9b + 19 = 0$

$\rightarrow \left(\frac{9}{2}\right)^2$

$b^2 + 9b + \left(\frac{9}{2}\right)^2 - \left(\frac{9}{2}\right)^2 + 19 = 0$

$\left[b^2 + 9b + \left(\frac{9}{2}\right)^2\right] - \frac{81}{4} + \frac{76}{4} = 0$

$\left(b + \frac{9}{2}\right)^2 - \frac{5}{4} = 0$

$\left(b + \frac{9}{2}\right)^2 = \frac{5}{4}$

$b + \frac{9}{2} = \pm\frac{\sqrt{5}}{2}$

$b = -\frac{9}{2} \pm \frac{\sqrt{5}}{2}$

$b = \frac{-9 \pm \sqrt{5}}{2}$

4 **a** $5x^2 - 2x - 4 = 0$

$$5\left(x^2 - \frac{2}{5}x - \frac{4}{5}\right) = 0$$

$$x^2 - \frac{2}{5}x - \frac{4}{5} = 0$$

$$\rightarrow \left(\frac{2}{10}\right)^2 = \left(\frac{1}{5}\right)^2$$

$$x^2 - \frac{2}{5}x + \left(\frac{1}{5}\right)^2 - \left(\frac{1}{5}\right)^2 - \frac{4}{5} = 0$$

$$\left[x^2 - \frac{2}{5}x + \left(\frac{1}{5}\right)^2\right] - \frac{1}{25} - \frac{20}{5} = 0$$

$$\left(x - \frac{1}{5}\right)^2 - \frac{21}{25} = 0$$

$$\left(x - \frac{1}{5}\right)^2 = \frac{21}{5}$$

$$x - \frac{1}{5} = \pm\frac{\sqrt{21}}{5}$$

$$x = \frac{1}{5} \pm \frac{\sqrt{21}}{2}$$

$$x = \frac{1 \pm \sqrt{21}}{5}$$

4 **b** $3a^2 + 2a - 4 = 0$

$$3\left(a^2 + \frac{2}{3}a - \frac{4}{3}\right) = 0$$

$$a^2 + \frac{2}{3}a - \frac{4}{3} = 0$$

$$\rightarrow \left(\frac{2}{6}\right)^2 = \left(\frac{1}{3}\right)^2$$

$$a^2 + \frac{2}{3}a + \left(\frac{1}{3}\right)^2 - \left(\frac{1}{3}\right)^2 - \frac{4}{3} = 0$$

$$\left[a^2 + \frac{2}{3}a + \left(\frac{1}{3}\right)^2\right] - \frac{1}{9} - \frac{12}{9} = 0$$

$$\left(a + \frac{1}{3}\right)^2 - \frac{13}{9} = 0$$

$$\left(a + \frac{1}{3}\right)^2 = \frac{13}{9}$$

$$a + \frac{1}{3} = \pm\frac{\sqrt{13}}{3}$$

$$a = -\frac{1}{3} \pm \frac{\sqrt{13}}{3}$$

$$a = \frac{-1 \pm \sqrt{13}}{3}$$

4 c $4b^2 - 2b = 5$

$$4b^2 - 2b - 5 = 0$$

$$4\left(b^2 - \frac{1}{2}b - \frac{5}{4}\right) = 0$$

$$b^2 - \frac{1}{2}b - \frac{5}{4} = 0$$

$$\rightarrow \left(\frac{1}{4}\right)^2$$

$$b^2 - \frac{1}{2}b + \left(\frac{1}{4}\right)^2 - \left(\frac{1}{4}\right)^2 - \frac{5}{4} = 0$$

$$\left[b^2 - \frac{1}{2}b + \left(\frac{1}{4}\right)^2\right] - \frac{1}{16} - \frac{20}{16} = 0$$

$$\left(b - \frac{1}{4}\right)^2 - \frac{21}{16} = 0$$

$$\left(b - \frac{1}{4}\right)^2 = \frac{21}{16}$$

$$b - \frac{1}{4} = \pm\frac{\sqrt{21}}{4}$$

$$b = \frac{1}{4} \pm \frac{\sqrt{21}}{4}$$

$$b = \frac{1 \pm \sqrt{21}}{4}$$

4 **d** $5x = 3x^2 - 3$

$3x^2 - 5x - 3 = 0$

$3\left(x^2 - \frac{5}{3}x - 1\right) = 0$

$x^2 - \frac{5}{3}x - 1 = 0$

$\rightarrow \left(\frac{5}{6}\right)^2$

$x^2 - \frac{5}{3}x + \left(\frac{5}{6}\right)^2 - \left(\frac{5}{6}\right)^2 - 1 = 0$

$\left[x^2 - \frac{5}{3}x + \left(\frac{5}{6}\right)^2\right] - \frac{25}{36} - \frac{36}{36} = 0$

$\left(x - \frac{5}{6}\right)^2 - \frac{61}{36} = 0$

$\left(x - \frac{5}{6}\right)^2 = \frac{61}{36}$

$x - \frac{5}{6} = \pm\frac{\sqrt{61}}{6}$

$x = \frac{5}{6} \pm \frac{\sqrt{61}}{6}$

$x = \frac{5 \pm \sqrt{61}}{6}$

4 **e** $4a - 3a^2 + 2 = 0$

$-3a^2 + 4a + 2 = 0$

$-3\left(a^2 - \frac{4}{3}a - \frac{2}{3}\right) = 0$

$a^2 - \frac{4}{3}a - \frac{2}{3} = 0$

$\rightarrow \left(\frac{4}{6}\right)^2 = \left(\frac{2}{3}\right)^2$

$a^2 - \frac{4}{3}a + \left(\frac{2}{3}\right)^2 - \left(\frac{2}{3}\right)^2 - \frac{2}{3} = 0$

$\left[a^2 - \frac{4}{3}a + \left(\frac{2}{3}\right)^2\right] - \frac{4}{9} - \frac{6}{9} = 0$

$\left(a - \frac{2}{3}\right)^2 - \frac{10}{9} = 0$

$\left(a - \frac{2}{3}\right)^2 = \frac{10}{9}$

$a - \frac{2}{3} = \pm\frac{\sqrt{10}}{3}$

$a = \frac{2}{3} \pm \frac{\sqrt{10}}{3}$

$a = \frac{2 \pm \sqrt{10}}{3}$

4 **f** $2b - 5b^2 = -1$

$$-5b^2 + 2b + 1 = 0$$

$$-5\left(b^2 - \frac{2}{5}b - \frac{1}{5}\right) = 0$$

$$b^2 - \frac{2}{5}b - \frac{1}{5} = 0$$

$$\rightarrow \left(\frac{2}{10}\right)^2 = \left(\frac{1}{5}\right)^2$$

$$b^2 - \frac{2}{5}b + \left(\frac{1}{5}\right)^2 - \left(\frac{1}{5}\right)^2 - \frac{1}{5} = 0$$

$$\left[b^2 - \frac{2}{5}b + \left(\frac{1}{5}\right)^2\right] - \frac{1}{25} - \frac{5}{25} = 0$$

$$\left(b - \frac{1}{5}\right)^2 - \frac{6}{26} = 0$$

$$\left(b - \frac{1}{5}\right)^2 = \frac{6}{25}$$

$$b - \frac{1}{5} = \pm\frac{\sqrt{6}}{5}$$

$$b = \frac{1}{5} \pm \frac{\sqrt{6}}{5}$$

$$b = \frac{1 \pm \sqrt{6}}{5}$$

4 **g** $-2 = 7x - 3x^2$

$3x^2 - 7x - 2 = 0$

$3\left(x^2 - \frac{7}{3}x - \frac{2}{3}\right) = 0$

$x^2 - \frac{7}{3}x - \frac{2}{3} = 0$

$\rightarrow \left(\frac{7}{6}\right)^2$

$x^2 - \frac{7}{3}x + \left(\frac{7}{6}\right)^2 - \left(\frac{7}{6}\right)^2 - \frac{2}{3} = 0$

$\left[x^2 - \frac{7}{3}x + \left(\frac{7}{6}\right)^2\right] - \frac{49}{36} - \frac{24}{36} = 0$

$\left(x - \frac{7}{6}\right)^2 - \frac{73}{36} = 0$

$\left(x - \frac{7}{6}\right)^2 = \frac{73}{36}$

$x - \frac{7}{6} = \pm\frac{\sqrt{73}}{6}$

$x = \frac{7}{6} \pm \frac{\sqrt{73}}{6}$

$x = \frac{7 \pm \sqrt{73}}{6}$

4 **h** $4a^2 - 4a = -7$

$$4a^2 - 4a + 7 = 0$$

$$4\left(a^2 - a + \frac{7}{4}\right) = 0$$

$$a^2 - a + \frac{7}{4} = 0$$

$$a^2 - a + \left(\frac{1}{2}\right)^2 - \left(\frac{1}{2}\right)^2 + \frac{7}{4} = 0$$

$$\left[a^2 - a + \left(\frac{1}{2}\right)^2\right] - \frac{1}{4} + \frac{7}{4} = 0$$

$$\left(a - \frac{1}{2}\right)^2 + \frac{6}{4} = 0$$

$\therefore$ *No Solution*

i $-3 - 5b^2 = -b$

$$-5b^2 + b - 3 = 0$$

$$-5\left(b^2 - \frac{1}{5}b + \frac{3}{5}\right) = 0$$

$$b^2 - \frac{1}{5}b + \frac{3}{5} = 0$$

$$b^2 - \frac{1}{5}b + \left(\frac{1}{10}\right)^2 - \left(\frac{1}{10}\right)^2 + \frac{3}{5} = 0$$

$$\left[b^2 - \frac{1}{5}b + \left(\frac{1}{10}\right)^2\right] - \frac{1}{100} + \frac{60}{100} = 0$$

$$\left(b - \frac{1}{10}\right)^2 + \frac{59}{100} = 0$$

$\therefore$ *No Solution*

Challenge Yourself

1 a

$$x^2 + 4x = \frac{29}{4}$$

$$x^2 + 4x - \frac{29}{4} = 0$$

$$x^2 + 4x + (2)^2 - (2)^2 - \frac{29}{4} = 0$$

$$[x^2 + 4x + (2)^2] - 4 - \frac{29}{4} = 0$$

$$(x+2)^2 - \frac{16}{4} - \frac{29}{4} = 0$$

$$(x+2)^2 - \frac{45}{4} = 0$$

$$(x+2)^2 = \frac{45}{4}$$

$$x + 2 = \pm\frac{\sqrt{45}}{2}$$

$$x = -2 \pm \frac{3\sqrt{5}}{2}$$

$$x = -\frac{4}{2} \pm \frac{3\sqrt{5}}{2}$$

$$x = \frac{-4 \pm 3\sqrt{5}}{2}$$

b

$$a^2 + 6a = -\frac{53}{9}$$

$$a^2 + 6a + \frac{53}{9} = 0$$

$$a^2 + 6a + (3)^2 - (3)^2 + \frac{53}{9} = 0$$

$$[a^2 + 6a + (3)^2] - 9 + \frac{53}{9} = 0$$

$$(a+3)^2 - \frac{81}{9} + \frac{53}{9} = 0$$

$$(a+3)^2 - \frac{28}{9} = 0$$

$$(a+3)^2 = \frac{28}{9}$$

$$a + 3 = \pm\frac{\sqrt{28}}{3}$$

$$a = -3 \pm \frac{2\sqrt{7}}{3}$$

$$a = -\frac{9}{3} \pm \frac{2\sqrt{7}}{3}$$

$$a = \frac{-9 \pm 2\sqrt{7}}{3}$$

1 c $$x^2 - 5x + \frac{1}{16} = 0$$

$$x^2 - 5x + \left(\frac{5}{2}\right)^2 - \left(\frac{5}{2}\right)^2 + \frac{1}{16} = 0$$

$$\left[x^2 - 5x + \left(\frac{5}{2}\right)^2\right] - \frac{25}{4} + \frac{1}{16} = 0$$

$$\left(x - \frac{5}{2}\right)^2 - \frac{100}{16} + \frac{1}{16} = 0$$

$$\left(x - \frac{5}{2}\right)^2 - \frac{99}{4} = 0$$

$$\left(x - \frac{5}{2}\right)^2 = \frac{99}{4}$$

$$x - \frac{5}{2} = \pm\frac{\sqrt{99}}{4}$$

$$x = \frac{5}{2} \pm \frac{3\sqrt{11}}{4}$$

$$x = \frac{10}{4} \pm \frac{3\sqrt{11}}{4}$$

$$x = \frac{10 \pm 3\sqrt{11}}{4}$$

1 **d**

$$3a^2 + 4a + \frac{10}{9} = 0$$

$$3\left(a^2 + \frac{4}{3}a + \frac{10}{27}\right) = 0$$

$$a^2 + \frac{4}{3}a + \frac{10}{27} = 0$$

$$\rightarrow \left(\frac{4}{6}\right)^2 = \left(\frac{2}{3}\right)^2$$

$$a^2 + \frac{4}{3}a + \left(\frac{2}{3}\right)^2 - \left(\frac{2}{3}\right)^2 + \frac{10}{27} = 0$$

$$\left[a^2 + \frac{4}{3}a + \left(\frac{2}{3}\right)^2\right] - \frac{4}{9} + \frac{10}{27} = 0$$

$$\left(x + \frac{2}{3}\right)^2 - \frac{12}{27} + \frac{10}{27} = 0$$

$$\left(x + \frac{2}{3}\right)^2 - \frac{2}{27} = 0$$

$$\left(x + \frac{2}{3}\right)^2 = \frac{2}{27}$$

$$x + \frac{2}{3} = \pm\frac{\sqrt{2}}{\sqrt{27}}$$

$$\frac{\sqrt{2}}{\sqrt{27}} = \frac{\sqrt{2}}{3\sqrt{3}} \times \frac{\sqrt{3}}{\sqrt{3}} = \frac{\sqrt{6}}{9}$$

$$x = -\frac{2}{3} \pm \frac{\sqrt{6}}{9}$$

$$x = -\frac{6}{9} \pm \frac{\sqrt{6}}{9}$$

$$x = \frac{-6 \pm \sqrt{6}}{9}$$

2 $x^2 + 6x - 2 = 0$

$x^2 + 6x + (3)^2 - (3)^2 - 2 = 0$

$[x^2 + 6x + (3)^2] - 9 - 2 = 0$

$(x + 3)^2 - 11 = 0$

$(x + 3)^2 - 11 = (x + p)^2 + q$

$\therefore p = 3 \ and \ q = -11$

3 $3(x + 2)^2 = 17$

$3(x^2 + 4x + 4) = 17$

$3x^2 + 12x + 12 = 17$

$3x^2 + 12x + 12 - 17 = 0$

$3x^2 + 12x - 5 = 0$

$3x^2 + 12x - 5 = 3x^2 + bx - 5 = 0$

$\therefore b = 12$

4 $4\left(x + \frac{1}{2}\right)^2 = 4$

$4\left(x^2 + x + \frac{1}{4}\right) = 4$

$4x^2 + 4x + 1 = 4$

$4x^2 + 4x + 1 - 4 = 0$

$4x^2 + 4x - 3 = 0$

$4x^2 + 4x - 3 = 4x^2 + 4x + m$

$\therefore m = -3$

5 $-3x^2 - 6x + 4 = 0$

$-3\left(x^2 + 2x - \frac{4}{3}\right) = 0$

$-3\left(x^2 + 2x + (1)^2 - (1)^2 - \frac{4}{3}\right) = 0$

$-3\left([x^2 + 2x + (1)^2] - 1 - \frac{4}{3}\right) = 0$

$-3\left((x + 1)^2 - \frac{3}{3} - \frac{4}{3}\right) = 0$

$-3\left((x + 1)^2 - \frac{7}{3}\right) = 0$

$-3(x + 1)^2 + 7 = 0$

$-3(x + 1)^2 + 7 = a(x + b)^2 + c$

$\therefore a = -3 \ and \ b = 1 \ and \ c = 7$

6 $2x^2 - 12x + 21 = 0$

$$2\left(x^2 - 6x + \frac{21}{2}\right) = 0$$

$$2\left(x^2 - 6x + (3)^2 - (3)^2 + \frac{21}{2}\right) = 0$$

$$2\left([x^2 - 6x + (3)^2] - 9 + \frac{21}{2}\right) = 0$$

$$2\left((x-3)^2 - \frac{18}{2} + \frac{21}{2}\right) = 0$$

$$2\left((x-3)^2 + \frac{3}{2}\right) = 0$$

$$2(x-3)^2 + 3 = 0$$

$$2(x-3)^2 + 3 = a(x+b)^2 + c$$

$$\therefore a = 2 \; and \; b = -3 \; and \; c = 3$$

Null Factor Law

Multiple Choice

1 **Option B**

$(x+2)(x-3)=0$

$x+2=0 \quad or\ x-3=0$

$x=-2 \qquad\qquad x=3$

$\therefore x=-2, x=3$

2 **Option D**

$x(x+7)=0$

$x=0 \quad or\ x+7=0$

$x=-7$

$\therefore x=0, x=-7$

3 **Option C**

$(3m-6)(m+3)=0$

$3m-6=0 \quad or \quad m+3=0$

$3m=6 \qquad\qquad m=-3$

$m=2$

$\therefore m=2, m=-3$

4 **Option A**

$x=4\ or\ x=-2$

$x-4=0 \quad or \quad x+2=0$

$\therefore (x-4)(x+2)=0$

5 **Option C**

$x=\frac{1}{3}\ or\ x=2$

$3x=1 \quad or \quad x-2=0$

$3x-1=0$

$\therefore (3x-1)(x-2)=0$

Time To Practice

1 **a** $x(x+3)=0$

$x=0 \quad or \quad x+3=0$

$x=-3$

$\therefore x=0, x=-3$

b $a(a+5)=0$

$a=0 \quad or \quad a+5=0$

$a=-5$

$\therefore a=0, a=-5$

c $b(b-7)=0$

$b=0 \quad or \quad b-7=0$

$b=7$

$\therefore b=0, b=7$

d $3x(x+2)=0$

$3x=0 \quad or \quad x+2=0$

$x=0 \qquad x=-2$

$\therefore x=0, x=-2$

e $4a(a-3)=0$

$4a=0 \quad or \quad a-3=0$

$a=0 \qquad a=3$

$\therefore a=0, a=3$

f $-b(b-9)=0$

$-b=0 \quad or \quad b-9=0$

$b=0 \qquad b=9$

$\therefore b=0, b=9$

g $-6x(4-x)=0$

$-6x=0 \quad or \quad 4-x=0$

$x=0 \qquad x=4$

$\therefore x=0, x=4$

h $-8a(1-a)=0$

$-8a=0 \quad or \quad 1-a=0$

$a=0 \qquad a=1$

$\therefore a=0, a=1$

i $-3b(-8-b)=0$

$-3b=0 \quad or \quad -8-b=0$

$b=0 \qquad b=-8$

$\therefore b=0, b=-8$

2 **a** $(x-5)(x+3)=0$

$x-5=0$ or $x+3=0$

$x=5$ $\quad x=-3$

$\therefore x=5, x=-3$

b $(a-4)(a+1)=0$

$a-4=0$ or $a+1=0$

$a=4$ $\quad a=-1$

$\therefore a=4, a=-1$

c $(b-2)(b-6)=0$

$b-2=0$ or $b-6=0$

$b=2$ $\quad b=6$

$\therefore b=2, b=6$

d $(4-x)(x-8)=0$

$4-x=0$ or $x-8=0$

$x=4$ $\quad x=8$

$\therefore x=4, x=8$

e $(a-8)(3-a)=0$

$a-8=0$ or $3-a=0$

$a=8$ $\quad a=3$

$\therefore a=8, a=3$

f $(b-3)(b-9)=0$

$b-3=0$ or $b-9=0$

$b=3$ $\quad b=9$

$\therefore b=3, b=9$

g $(x-4)(x+3)=0$

$x-4=0$ or $x+3=0$

$x=4$ $\quad x=-3$

$\therefore x=4, x=-3$

h $(2-a)(9-a)=0$

$2-a=0$ or $9-a=0$

$a=2$ $\quad a=9$

$\therefore a=2, a=9$

i $(5-b)(6-b)=0$

$5-b=0$ or $6-b=0$

$b=5$ $\quad b=6$

$\therefore b=5, b=6$

3 **a** $2(3x+1)(x-4)=0$

$(3x+1)(x-4)=0$

$3x+1=0 \quad or \quad x-4=0$

$3x=-1 \qquad x=4$

$x=-\frac{1}{3}$

$\therefore x=-\frac{1}{3}, x=4$

b $-3(2m-5)(m+6)=0$

$(2m-5)(m+6)=0$

$2m-5=0 \quad or \quad m+6=0$

$2m=5 \qquad m=-6$

$m=\frac{5}{2}$

$\therefore m=\frac{5}{2}, m=-6$

c $4(p-2)(5p-3)=0$

$(p-2)(5p-3)=0$

$p-2=0 \quad or \quad 5p-3=0$

$p=2 \qquad 5p=3$

$p=\frac{3}{5}$

$\therefore p=2, p=\frac{3}{5}$

d $-2(4a+1)(2a-7)=0$

$(4a+1)(2a-7)=0$

$4a+1=0 \quad or \quad 2a-7=0$

$4a=-1 \qquad 2a=7$

$a=-\frac{1}{4} \qquad a=\frac{7}{2}$

$\therefore a=-\frac{1}{4}, a=\frac{7}{2}$

e $3(2u+5)(u-1)=0$

$(2u+5)(u-1)=0$

$2u+5=0 \quad or \quad u-1=0$

$2u=-5 \qquad u=1$

$u=-\frac{5}{2}$

$\therefore u=-\frac{5}{2}, u=1$

f $-5(w-3)(3w+4)=0$

$(w-3)(3w+4)=0$

$w-3=0 \quad or \quad 3w+4=0$

$w=3 \qquad 3w=-4$

$w=-\frac{4}{3}$

$\therefore w=3, w=-\frac{4}{3}$

3 **g** $-7(5-3x)(3x+2)=0$

$(5-3x)(3x+2)=0$

$5-3x=0 \quad or \quad 3x+2=0$

$3x=5 \qquad 3x=-2$

$x=\frac{5}{3} \qquad x=-\frac{2}{3}$

$\therefore x=\frac{5}{3}, x=-\frac{2}{3}$

h $-6(7-2y)(4-3y)=0$

$(7-2y)(4-3y)=0$

$7-2y=0 \quad or \quad 4-3y=0$

$2y=7 \qquad 3y=4$

$y=\frac{7}{2} \qquad y=\frac{4}{3}$

$\therefore y=\frac{7}{2}, y=\frac{4}{3}$

i $-9(4w+3)(2w-9)=0$

$(4w+3)(2w-9)=0$

$4w+3=0 \quad or \quad 2w-9=0$

$4w=-3 \qquad 2w=9$

$w=-\frac{3}{4} \qquad w=\frac{9}{2}$

$\therefore w=-\frac{3}{4}, w=\frac{9}{2}$

j $8(1-7t)(2t-11)=0$

$(1-7t)(2t-11)=0$

$1-7t=0 \quad or \quad 2t-11=0$

$7t=1 \qquad 2t=11$

$t=\frac{1}{7} \qquad t=\frac{11}{2}$

$\therefore t=\frac{1}{7}, t=\frac{11}{2}$

Challenge Yourself

1 $3(x+2)(x-p)=0$

$(x+2)(x-p)=0$

$x+2=0 \quad or \quad x-p=0$

$x=-2 \qquad x=p$

$x=5$ *is one solution*,

$\therefore p=5$

2 $(kx+3)(x-1)=0$

$kx+3=0 \quad or \quad x-1=0$

$kx=-3 \qquad x=1$

$x=-\frac{3}{k}$

$x=-\frac{3}{2}$ *is one solution*,

$\therefore -\frac{3}{k}=-\frac{3}{2}$

$k=2$

3

$(x+5)(a-3x)=0$

$x+5=0 \quad or \quad a-3x=0$

$x=-5 \qquad x=\frac{a}{3}$

$x=\frac{1}{3}$ *is one solution ,*

$\therefore \frac{a}{3}=\frac{1}{3}$

$a=1$

4

$x=-12, x=9$

$x+12=0 \quad or \quad x-9=0$

$(x+12)(x-9)=0$

$x^2-9x+12x-108=0$

$x^2+3x-108=0$

$x^2+3x-108=ax^2+bx+c$

$\therefore a=1, b=3, c=-108$

5

$x=\frac{1}{4}, x=-3$

$4x=1 \quad or \quad x+3=0$

$4x-1=0$

$(4x-1)(x+3)=0$

$4x^2+12x-x-3=0$

$4x^2+11x-3=0$

$4x^2+11x-3=ax^2+bx+c$

$\therefore a=4, b=11, c=-3$

6

$x=\frac{2}{3}, x=\frac{1}{2}$

$3x=2 \quad or \quad 2x=1$

$3x-2=0 \qquad 2x-1=0$

$(3x-2)(2x-1)=0$

$6x^2-3x-4x+2=0$

$6x^2-7x+2=0$

$6x^2-7x+2=ax^2+bx+c$

$\therefore a=6, b=-7, c=2$

Factorising

Multiple Choice

1 **Option D**

$4x^2 + 12x = 0$

$4x(x + 3) = 0$

$4x = 0 \quad or \quad x + 3 = 0$

$x = 0 \qquad x = -3$

$\therefore x = 0, x = -3$

2 **Option A**

$x^2 + 10x + 25 = 0$

$(x + 5)^2 = 0$

$x + 5 = 0$

$x = -5$

$\therefore x = -5$

3 **Option B**

$x^2 - 10x + 16 = 0$

$(x - 8)(x - 2) = 0$

$x - 8 = 0 \quad or \quad x - 2 = 0$

$x = 8 \qquad x = 2$

$\therefore x = 8, x = 2$

4 **Option D**

$x^2 + 3x - 4 = 0$

$(x - 1)(x + 4) = 0$

$x - 1 = 0 \quad or \quad x + 4 = 0$

$x = 1 \qquad x = -4$

5 **Option A**

$x^2 - 45 = 4x$

$x^2 - 4x - 45 = 0$

$(x - 9)(x + 5) = 0$

$x - 9 = 0 \quad or \quad x + 5 = 0$

$x = 9 \qquad x = -5$

Time To Practice

1 a

$x^2 + 7x = 0$

$x(x + 7) = 0$

$x = 0 \quad or \quad x + 7 = 0$

$x = 0 \qquad x = -7$

$\therefore x = 0, x = -7$

b

$a^2 - 5a = 0$

$a(a - 5) = 0$

$a = 0 \quad or \quad a - 5 = 0$

$a = 5$

$\therefore a = 0, a = 5$

c

$b^2 - 10b = 0$

$b(b - 10) = 0$

$b = 0 \quad or \quad b - 10 = 0$

$b = 10$

$\therefore b = 0, b = 10$

d

$3x^2 - 9x = 0$

$3x(x - 3) = 0$

$3x = 0 \quad or \quad x - 3 = 0$

$x = 0 \qquad x = 3$

$\therefore x = 0, x = 3$

e

$5a^2 - 10a = 0$

$5a(a - 2) = 0$

$5a = 0 \quad or \quad a - 2 = 0$

$a = 0 \qquad a = 2$

$\therefore a = 0, a = 2$

f

$-b^2 - b = 0$

$-b(b + 1) = 0$

$-b = 0 \quad or \quad b + 1 = 0$

$b = 0 \qquad b = -1$

$\therefore b = 0, b = -1$

g

$36x = 6x^2$

$6x^2 - 36x = 0$

$6x(x - 6) = 0$

$6x = 0 \quad or \quad x - 6 = 0$

$x = 0 \qquad x = 6$

$\therefore x = 0, x = 6$

h

$2a^2 = 22a$

$2a^2 - 22a = 0$

$2a(a - 11) = 0$

$2a = 0 \quad or \quad a - 11 = 0$

$a = 0 \qquad a = 11$

$\therefore a = 0, a = 11$

1 **i** $-8b^2 = 24b$

$24b + 8b^2 = 0$

$8b(3 + b) = 0$

$8b = 0 \quad or \quad 3 + b = 0$

$b = 0 \qquad\qquad b = -3$

$\therefore b = 0, b = -3$

2 **a** $x^2 + 5x + 6 = 0$

$(x + 2)(x + 3) = 0$

$x + 2 = 0 \quad or \quad x + 3 = 0$

$x = -2 \qquad\qquad x = -3$

$\therefore x = -2, x = -3$

b $a^2 - 7a + 10 = 0$

$(a - 5)(a - 2) = 0$

$a - 5 = 0 \quad or \quad a - 2 = 0$

$a = 5 \qquad\qquad a = 2$

$\therefore a = 5, a = 2$

c $2b^2 + 8b + 6 = 0$

$2(b^2 + 4b + 3) = 0$

$2(b + 1)(b + 3) = 0$

$(b + 1)(b + 3) = 0$

$b + 1 = 0 \quad or \quad b + 3 = 0$

$b = -1 \qquad\qquad b = -3$

$\therefore b = -1, b = -3$

d $2x^2 - 4x = 6$

$2x^2 - 4x - 6 = 0$

$2(x^2 - 2x - 3) = 0$

$(x - 3)(x + 1) = 0$

$x - 3 = 0 \quad or \quad x + 1 = 0$

$x = 3 \qquad\qquad x = -1$

$\therefore x = 3, x = -1$

2 **e**

$3a^2 = 11a - 10$

$3a^2 - 11a + 10 = 0$

$(3a - 5)(a - 2) = 0$

$3a - 5 = 0 \quad or \quad a - 2 = 0$

$3a = 5 \qquad\qquad a = 2$

$a = \frac{5}{3}$

$\therefore a = \frac{5}{3}, a = 2$

f

$7b - 2b^2 = 3$

$2b^2 - 7b + 3 = 0$

$(2b - 1)(b - 3) = 0$

$2b - 1 = 0 \quad or \quad b - 3 = 0$

$b = 1 \qquad\qquad b = 3$

$b = \frac{1}{2}$

$\therefore b = \frac{1}{2}, b = 3$

g

$10x - 3x^2 = 8$

$3x^2 - 10x + 8 = 0$

$(3x - 4)(x - 2) = 0$

$3x - 4 = 0 \quad or \quad x - 2 = 0$

$3x = 4 \qquad\qquad x = 2$

$x = \frac{4}{3}$

$\therefore x = \frac{4}{3}, x = 2$

h

$4a^2 = 3 + 4a$

$4a^2 - 4a - 3 = 0$

$(2a + 1)(2a - 3) = 0$

$2a + 1 = 0 \quad or \quad 2a - 3 = 0$

$2a = -1 \qquad\qquad 2a = 3$

$a = -\frac{1}{2} \qquad\qquad a = \frac{3}{2}$

$\therefore a = -\frac{1}{2}, a = \frac{3}{2}$

i

$15 = 11b - 2b^2$

$2b^2 - 11b + 15 = 0$

$(2b - 5)(b - 3) = 0$

$2b - 5 = 0 \quad or \quad b - 3 = 0$

$2b = 5 \qquad\qquad b = 3$

$b = \frac{5}{2}$

$\therefore b = \frac{5}{2}, b = 3$

3 **a** $4x^2 - 5(4x - 5) = 0$

$4x^2 - 20x + 25 = 0$

$(2x)^2 - 20x + 5^2 = 0$

$(2x - 5)^2 = 0$

$2x - 5 = 0$

$2x = 5$

$x = \frac{5}{2}$

b $4p^2 + 3(4p + 3) = 0$

$4p^2 + 12p + 9$

$(2p)^2 + 12p + 3^2 = 0$

$(2p + 3)^2 = 0$

$2p + 3 = 0$

$2p = -3$

$p = -\frac{3}{2}$

c $x(x + 4) = 4x + 9$

$x^2 + 4x = 4x + 9$

$x^2 + 4x - 4x = 9$

$x^2 = 9$

$x = \pm\sqrt{9}$

$x = \pm 3$

d $4(u^2 + 2) - 9 = 0$

$4u^2 + 8 - 9 = 0$

$4u^2 - 1 = 0$

$4u^2 - 1 = 0$

$4u^2 = 1$

$u^2 = \frac{1}{4}$

$u = \pm\sqrt{\frac{1}{4}} = \pm\frac{1}{2}$

e $s^2 + s - 9 = 5s - 4$

$s^2 + s - 9 - 5s + 4 = 0$

$s^2 - 4s - 5 = 0$

$(s - 5)(s + 1) = 0$

$s - 5 = 0 \quad or \quad s + 1 = 0$

$s = 5 \qquad\qquad s = -1$

$\therefore s = 5, s = -1$

f $3y^2 - 4(2y - 1) = 0$

$3y^2 - 8y + 4 = 0$

$(3y - 2)(y - 2) = 0$

$3y - 2 = 0 \quad or \quad y - 2 = 0$

$3y = 2 \qquad\qquad y = 2$

$y = \frac{2}{3}$

$\therefore y = \frac{2}{3}, y = 2$

3 **g**

$3(1 - 2a^2) - 17a = 0$

$3 - 6a^2 - 17a = 0$

$6a^2 + 17a - 3 = 0$

$(6a - 1)(a + 3) = 0$

$6a - 1 = 0 \quad or \quad a + 3 = 0$

$6a = 1 \qquad\qquad a = -3$

$a = \frac{1}{6}$

$\therefore a = \frac{1}{6}, a = -3$

h

$9x^2 - (x - 1)^2 = 0$

$9x^2 - (x^2 - 2x + 1) = 0$

$9x^2 - x^2 + 2x - 1 = 0$

$8x^2 + 2x - 1 = 0$

$(4x - 1)(2x + 1) = 0$

$4x - 1 = 0 \quad or \quad 2x + 1 = 0$

$4x = 1 \qquad\qquad 2x = -1$

$x = \frac{1}{4} \qquad\qquad x = -\frac{1}{2}$

$\therefore x = \frac{1}{4}, x = \frac{1}{2}$

i

$3t(t + 6) = 4(t - 2)$

$3t^2 + 18t = 4t - 8$

$3t^2 + 18t - 4t + 8 = 0$

$3t^2 + 14t + 8 = 0$

$(3t + 2)(t + 4) = 0$

$3t + 2 = 0 \quad or \quad t + 4 = 0$

$3t = -2 \qquad\qquad t = -4$

$t = -\frac{2}{3}$

$\therefore t = -\frac{2}{3}, t = -4$

j

$6k(k + 1) = 15 - 4k + k^2$

$6k^2 + 6k = 15 - 4k + k^2$

$6k^2 + 6k - 15 + 4k - k^2 = 0$

$5k^2 + 10k - 15 = 0$

$5(k^2 + 2k - 3) = 0$

$5(k - 1)(k + 3) = 0$

$(k - 1)(k + 3) = 0$

$k - 1 = 0 \quad or \quad k + 3 = 0$

$k = 1 \qquad\qquad k = -3$

$\therefore k = 1, k = -3$

4 **a** $\frac{5x + 84}{x} = x$

$5x + 84 = x^2$

$x^2 - 5x - 84 = 0$

$(x - 12)(x + 7) = 0$

$x - 12 = 0 \quad or \quad x + 7 = 0$

$x = 12 \qquad\qquad x = 7$

$\therefore x = 12, x = -7$

b $\frac{3}{v} = v + 2$

$3 = v(v + 2)$

$3 = v^2 + 2v$

$v^2 + 2v - 3 = 0$

$(v - 1)(v + 3) = 0$

$v - 1 = 0 \quad or \quad v + 3 = 0$

$v = 1 \qquad\qquad v = -3$

$\therefore v = 1, v = -3$

c $\frac{4}{x - 2} = x + 1$

$4 = (x - 2)(x + 1)$

$4 = x^2 + x - 2x - 2$

$4 = x^2 - x - 2$

$x^2 - x - 6 = 0$

$(x - 3)(x + 2) = 0$

$x - 3 = 0 \quad or \quad x + 2 = 0$

$x = 3 \qquad\qquad x = -2$

$\therefore x = 3, x = -2$

d $\frac{1}{h} = 3 - 2h$

$1 = h(3 - 2h)$

$1 = 3h - 2h^2$

$2h^2 - 3h + 1 = 0$

$(2h - 1)(h - 1) = 0$

$2h - 1 = 0 \quad or \quad h - 1 = 0$

$2h = 1 \qquad\qquad h = 1$

$h = \frac{1}{2}$

$\therefore h = \frac{1}{2}, h = 1$

4 e

$$\frac{20-3a}{a}=2a$$
$$20-3a=2a^2$$
$$2a^2+3a-20=0$$
$$(2a-5)(a+4)=0$$

$2a-5=0$ *or* $a+4=0$

$2a=5$ $\quad a=-4$

$a=\frac{5}{2}$

$\therefore a=\frac{5}{2}, a=-4$

f

$$\frac{7w+10}{2w}=3w$$
$$7w+10=6w^2$$
$$6w^2-7w-10=0$$
$$(6w+5)(w-2)=0$$

$6w+5=0$ *or* $w-2=0$

$6w=-5$ $\quad w=2$

$w=-\frac{5}{6}$

$\therefore w=-\frac{5}{6}, w=2$

Challenge Yourself

1 a

$$12x=13-\frac{3}{x}$$
$$12x=\frac{13}{1}-\frac{3}{x}$$
$$12x=\frac{13x-3}{x}$$
$$12x^2=13x-3$$
$$12x^2-13x+3=0$$
$$(3x-1)(4x-3)=0$$

$3x-1=0$ *or* $4x-3=0$

$3x=1$ $\quad 4x=3$

$x=\frac{1}{3}$ $\quad x=\frac{3}{4}$

$\therefore x=\frac{1}{3}, x=\frac{3}{4}$

b

$$\frac{10}{x}-21=10x$$
$$\frac{10}{x}-\frac{21}{1}=10x$$
$$\frac{10-21x}{x}=10x$$
$$10-21x=10x^2$$
$$10x^2+21x-10=0$$
$$(5x-2)(2x+5)=0$$

$5x-2=0$ *or* $2x+5=0$

$5x=2$ $\quad 2x=-5$

$x=\frac{2}{5}$ $\quad x=-\frac{5}{2}$

$\therefore x=\frac{2}{5}, x=-\frac{5}{2}$

1 c $\frac{x+4}{2}-\frac{3}{x-3}=1$

$$\frac{(x+4)(x-3)-6}{2(x-3)}=1$$

$$(x+4)(x-3)-6=2(x-3)$$

$$x^2-3x+4x-12-6=2x-6$$

$$x^2+x-18=2x-6$$

$$x^2+x-18-2x+6=0$$

$$x^2-x-12=0$$

$$(x-4)(x+3)=0$$

$x-4=0$ *or* $x+3=0$

$x=4$ $\quad x=-3$

$\therefore x=4, x=-3$

1 d $\frac{x}{x-2} - \frac{x+1}{x+4} = 1$

$\frac{x(x+4) - (x+1)(x-2)}{(x-2)(x+4)} = 1$

$x(x+4) - (x+1)(x-2) = (x-2)(x+4)$

$x^2 + 4x - (x^2 - 2x + x - 2) = x^2 + 4x - 2x - 8$

$x^2 + 4x - (x^2 - x - 2) = x^2 + 2x - 8$

$x^2 + 4x - x^2 + x + 2 = x^2 + 2x - 8$

$5x + 2 = x^2 + 2x - 8$

$x^2 + 2x - 8 - 5x - 2 = 0$

$x^2 - 3x - 10 = 0$

$(x-5)(x+2) = 0$

$x - 5 = 0 \quad or \quad x + 2 = 0$

$x = 5 \qquad\qquad x = -2$

$\therefore x = 5, x = -2$

2 $x = -7, x = p$

$x + 7 = 0 \quad or \quad x - p = 0$

$(x+7)(x-p) = 0$

$\therefore (x-7)(x-p) = x^2 - x - 56$

$7 \times -p = -56$

$-p = -8$

$p = 8$

3 $(x-a)(x-b)=0$

$(x-2)(x-b)=0$

$(x-2)(x-b)=x^2-px+10$

$-2\times -b=10$

$2b=10$

$b=5$

$(x-2)(x-5)=0$

$x^2-5x-2x+10=0$

$x^2-7x+10=0$

$x^2-7x+10=x^2-px+10$

$p=-7$

$\therefore b=5 \; and \; p=-7$

4 For the equation to have exactly one solution, it must be a **perfect square trinomial**

$x^2+10x+k=0$

$\rightarrow \left(\frac{10}{2}\right)^2=(5)^2$

$x^2+10x+(5)^2=0$

$x^2+10x+25=0$

$\therefore k=25$

$x^2+10x+(5)^2=0$

$(x+5)^2=0$

$x+5=0$

$x=-5$

$\therefore k=25 \; and \; x=-5$

5 $ax^2 + 9x + 2 = 0$

Sub $x = -2$ into equation.

$a(-2)^2 + 9(-2) + 2 = 0$

$4a - 18 + 2 = 0$

$4a - 16 = 0$

$4a = 16$

$a = 4$

6 $3x^2 + 7x + c = 0$

Sub $x = -3$ into equation.

$3(-3)^2 + 7(-3) + c = 0$

$27 - 21 + c = 0$

$6 + c = 0$

$c = -6$

The Quadratic Formula

Multiple Choice

1 **Option C**

If $b^2 - 4ac < 0$

No or 0 real solutions

2 **Option B**

$a = 1, b = -3, c = 2$

$\Delta = b^2 - 4ac$

$= (-3)^2 - 4(1)(2)$

$= 9 - 8 = 1$

3 **Option A**

$3x^2 - 6x + 2 = 0$

$a = 3, b = -6, c = 2$

$\Delta = b^2 - 4ac$

$= (-6)^2 - 4(3)(2)$

$= 36 - 24$

$= 12$

4 **Option D**

$-2x^2 = 7x - 1$

$-2x^2 - 7x + 1 = 0$

$\therefore a = -2, b = -7, c = 1$

5 **Option B**

To use the quadratic formula there must be an x^2 term. In other words the coefficient of the x^2 term cannot be zero. i.e. $a \neq 0$.

$\therefore x + 5 = 0$

Time To Practice

1 **a** $x^2 + 5x + 3 = 0$

$a = 1, b = 5, c = 3$

$\Delta = b^2 - 4ac$

$= 5^2 - 4(1)(3)$

$= 25 - 12$

$= 13$

b $x^2 + 6x + 9 = 0$

$a = 1, b = 6, c = 9$

$\Delta = b^2 - 4ac$

$= 6^2 - 4(1)(9)$

$= 36 - 36$

$= 0$

c $2x^2 - 3x + 4 = 0$

$a = 2, b = -3, c = 4$

$\Delta = b^2 - 4ac$

$= (-3)^2 - 4(2)(4)$

$= 9 - 32$

$= -23$

d $3x^2 + 2x + 1 = 0$

$a = 3, b = 2, c = 1$

$\Delta = b^2 - 4ac$

$= 2^2 - 4(3)(1)$

$= 4 - 12$

$= -8$

e $5x^2 + 3x - 2 = 0$

$a = 5, b = 3, c = -2$

$\Delta = b^2 - 4ac$

$= 3^2 - 4(5)(-2)$

$= 9 + 40$

$= 49$

f $-3x^2 + 4x + 5 = 0$

$a = -3, b = 4, c = 5$

$\Delta = b^2 - 4ac$

$= 4^2 - 4(-3)(5)$

$= 16 + 60$

$= 76$

2 a $x^2 + 7x - 1 = 0$

$a = 1, b = 7, c = -1$

$\Delta = b^2 - 4ac$

$= 7^2 - 4(1)(-1)$

$= 49 + 4$

$= 53$

Since $\Delta > 0$, there are TWO real solutions.

b $x^2 + 4x + 5 = 0$

$a = 1, b = 4, c = 5$

$\Delta = b^2 - 4ac$

$= 4^2 - 4(1)(5)$

$= 16 - 20$

$= -4$

Since $\Delta < 0$, there are NO real solutions.

c $2x^2 - 6x + 3 = 0$

$a = 2, b = -6, c = 3$

$\Delta = b^2 - 4ac$

$= (-6)^2 - 4(2)(3)$

$= 36 - 24$

$= 12$

Since $\Delta > 0$, there are TWO real solutions.

d $-2x^2 + 3x - 1 = 0$

$a = -2, b = 3, c = -1$

$\Delta = b^2 - 4ac$

$= 3^2 - 4(-2)(-1)$

$= 9 - 8$

$= 1$

Since $\Delta > 0$, there are TWO real solutions.

e $4x^2 - 5x + 2 = 0$

$a = 4, b = -5, c = 2$

$\Delta = b^2 - 4ac$

$= (-5)^2 - 4(4)(2)$

$= 25 - 32$

$= -7$

Since $\Delta < 0$, there are NO real solutions.

f $9x^2 + 6x + 1 = 0$

$a = 9, b = 6, c = 1$

$\Delta = b^2 - 4ac$

$= 6^2 - 4(9)(1)$

$= 36 - 36$

$= 0$

Since $\Delta = 0$, there is ONE real solution.

3 **a** $x^2 + 5x - 6 = 0$

$a = 1, b = 5, c = -6$

$$x = \frac{-5 \pm \sqrt{5^2 - 4(1)(-6)}}{2(1)}$$

$$x = \frac{-5 \pm \sqrt{25 + 24}}{2}$$

$$x = \frac{-5 \pm \sqrt{49}}{2} = \frac{-5 \pm 7}{2}$$

$$x = \frac{-5 - 7}{2} \quad or \quad x = \frac{-5 + 7}{2}$$

$$x = \frac{-12}{2} \qquad x = \frac{2}{2}$$

$x = -6 \quad or \quad x = 1$

b $x^2 - 6x + 8 = 0$

$a = 1, b = -6, c = 8$

$$x = \frac{-(-6) \pm \sqrt{(-6)^2 - 4(1)(8)}}{2(1)}$$

$$x = \frac{6 \pm \sqrt{36 - 32}}{2}$$

$$x = \frac{6 \pm \sqrt{4}}{2} = \frac{6 \pm 2}{2}$$

$$x = \frac{6 - 2}{2} \quad or \quad x = \frac{6 + 2}{2}$$

$$x = \frac{4}{2} \qquad x = \frac{8}{2}$$

$x = 2 \quad or \quad x = 4$

c $x^2 - 8x + 16 = 0$

$a = 1, b = -8, c = 16$

$$x = \frac{-(-8) \pm \sqrt{(-8)^2 - 4(1)(16)}}{2(1)}$$

$$x = \frac{8 \pm \sqrt{64 - 64}}{2}$$

$$x = \frac{8 \pm \sqrt{0}}{2}$$

$$x = \frac{8}{2}$$

$x = 4$

d $-x^2 + 5x + 14 = 0$

$a = -1, b = 5, c = 14$

$$x = \frac{-5 \pm \sqrt{5^2 - 4(-1)(14)}}{2(-1)}$$

$$x = \frac{-5 \pm \sqrt{25 + 56}}{-2}$$

$$x = \frac{-5 \pm \sqrt{81}}{-2} = \frac{-5 \pm 9}{-2}$$

$$x = \frac{-5 - 9}{-2} \quad or \quad x = \frac{-5 + 9}{-2}$$

$$x = \frac{-14}{-2} \qquad x = \frac{4}{-2}$$

$x = 7 \quad or \quad x = -2$

3 **e** $-x^2 + 8x + 9 = 0$

$a = -1, b = 8, c = 9$

$$x = \frac{-8 \pm \sqrt{8^2 - 4(-1)(9)}}{2(-1)}$$

$$x = \frac{-8 \pm \sqrt{64 + 36}}{-2}$$

$$x = \frac{-8 \pm \sqrt{100}}{-2} = \frac{-8 \pm 10}{-2}$$

$$x = \frac{-8 - 10}{-2} \quad or \quad x = \frac{-8 + 10}{-2}$$

$$x = \frac{-18}{-2} \qquad x = \frac{2}{-2}$$

$x = 9 \quad or \quad x = -1$

f $x^2 - 12x + 36 = 0$

$a = 1, b = -12, c = 36$

$$x = \frac{-(-12) \pm \sqrt{(-12)^2 - 4(1)(36)}}{2(1)}$$

$$x = \frac{12 \pm \sqrt{144 - 144}}{2}$$

$$x = \frac{12 \pm \sqrt{0}}{2}$$

$$x = \frac{12}{2}$$

$x = 6$

g $-x^2 + 11x - 30 = 0$

$a = -1, b = 11, c = -30$

$$x = \frac{-11 \pm \sqrt{11^2 - 4(-1)(-30)}}{2(-1)}$$

$$x = \frac{-11 \pm \sqrt{121 - 120}}{-2}$$

$$x = \frac{-11 \pm \sqrt{1}}{-2} = \frac{11 \pm 1}{-2}$$

$$x = \frac{-11 - 1}{-2} \quad or \quad x = \frac{-11 + 1}{-2}$$

$$x = \frac{-12}{-2} \qquad x = \frac{-10}{-2}$$

$x = 6 \quad or \quad x = 5$

h $25x^2 + 20x + 4 = 0$

$a = 25, b = 20, c = 4$

$$x = \frac{-20 \pm \sqrt{20^2 - 4(25)(4)}}{2(25)}$$

$$x = \frac{-20 \pm \sqrt{400 - 400}}{50}$$

$$x = \frac{-20 \pm \sqrt{0}}{50}$$

$$x = \frac{-20}{50}$$

$$x = -\frac{2}{5}$$

3 **i** $4x^2 - 4x + 1 = 0$

$a = 4, b = -4, c = 1$

$$x = \frac{-(-4) \pm \sqrt{(-4)^2 - 4(4)(1)}}{2(4)}$$

$$x = \frac{4 \pm \sqrt{16 - 16}}{8}$$

$$x = \frac{4 \pm \sqrt{0}}{8}$$

$$x = \frac{4}{8}$$

$$x = \frac{1}{2}$$

4 **a** $2x^2 + x - 3 = 0$

$a = 2, b = 1, c = -3$

$$x = \frac{-1 \pm \sqrt{1^2 - 4(2)(-3)}}{2(2)}$$

$$x = \frac{-1 \pm \sqrt{1 + 24}}{4}$$

$$x = \frac{-1 \pm \sqrt{25}}{4} = \frac{-1 \pm 5}{4}$$

$$x = \frac{-1 - 5}{4} \quad or \quad x = \frac{-1 + 5}{4}$$

$$x = \frac{-6}{4} \qquad x = \frac{4}{4}$$

$$x = -\frac{3}{2} \quad or \quad x = 1$$

b $2x^2 - 9x + 4 = 0$

$a = 2, b = -9, c = 4$

$$x = \frac{-(-9) \pm \sqrt{(-9)^2 - 4(2)(4)}}{2(2)}$$

$$x = \frac{9 \pm \sqrt{81 - 32}}{4}$$

$$x = \frac{9 \pm \sqrt{49}}{4} = \frac{9 \pm 7}{4}$$

$$x = \frac{9 - 7}{4} \quad or \quad x = \frac{9 + 7}{4}$$

$$x = \frac{2}{4} \qquad x = \frac{16}{4}$$

$$x = \frac{1}{2} \quad or \quad x = 4$$

4 **c** $2x^2 + 5x + 3 = 0$

$a = 2, b = 5, c = 3$

$$x = \frac{-5 \pm \sqrt{5^2 - 4(2)(3)}}{2(2)}$$

$$x = \frac{-5 \pm \sqrt{25 - 24}}{4}$$

$$x = \frac{-5 \pm \sqrt{1}}{4} = \frac{-5 \pm 1}{4}$$

$$x = \frac{-5 - 1}{4} \quad or \quad x = \frac{-5 + 1}{4}$$

$$x = \frac{-6}{4} \qquad\qquad x = \frac{-4}{4}$$

$$x = -\frac{3}{2} \quad or \quad x = -1$$

d $4x^2 - 8x + 3 = 0$

$a = 4, b = -8, c = 3$

$$x = \frac{-(-8) \pm \sqrt{(-8)^2 - 4(4)(3)}}{2(4)}$$

$$x = \frac{8 \pm \sqrt{64 - 48}}{8}$$

$$x = \frac{8 \pm \sqrt{16}}{8} = \frac{8 \pm 4}{8}$$

$$x = \frac{8 - 4}{8} \quad or \quad x = \frac{8 + 4}{8}$$

$$x = \frac{4}{8} \qquad\qquad x = \frac{12}{8}$$

$$x = \frac{1}{2} \quad or \quad x = \frac{3}{2}$$

e $12x^2 - 7x + 1 = 0$

$a = 12, b = -7, c = 1$

$$x = \frac{-(-7) \pm \sqrt{(-7)^2 - 4(12)(1)}}{2(12)}$$

$$x = \frac{7 \pm \sqrt{49 - 48}}{24}$$

$$x = \frac{7 \pm \sqrt{1}}{24} = \frac{7 \pm 1}{24}$$

$$x = \frac{7 - 1}{24} \quad or \quad x = \frac{7 + 1}{24}$$

$$x = \frac{6}{24} \qquad\qquad x = \frac{8}{24}$$

$$x = \frac{1}{4} \quad or \quad x = \frac{1}{3}$$

f $9x^2 - 9x + 2 = 0$

$a = 9, b = -9, c = 2$

$$x = \frac{-(-9) \pm \sqrt{(-9)^2 - 4(9)(2)}}{2(9)}$$

$$x = \frac{9 \pm \sqrt{81 - 72}}{18}$$

$$x = \frac{9 \pm \sqrt{9}}{18} = \frac{9 \pm 3}{18}$$

$$x = \frac{9 - 3}{18} \quad or \quad x = \frac{9 + 3}{18}$$

$$x = \frac{6}{18} \qquad\qquad x = \frac{12}{18}$$

$$x = \frac{1}{3} \quad or \quad x = \frac{2}{3}$$

4 **g** $3x^2 - x - 10 = 0$

$a = 3, b = -1, c = -10$

$$x = \frac{-(-1) \pm \sqrt{(-1)^2 - 4(3)(-10)}}{2(3)}$$

$$x = \frac{1 \pm \sqrt{1 + 120}}{6}$$

$$x = \frac{1 \pm \sqrt{121}}{6} = \frac{1 \pm 11}{6}$$

$$x = \frac{1 - 11}{6} \quad or \quad x = \frac{1 + 11}{6}$$

$$x = \frac{-10}{6} \qquad x = \frac{12}{6}$$

$$x = -\frac{5}{3} \quad or \quad x = 2$$

h $12x^2 + 4x - 1 = 0$

$a = 12, b = 4, c = -1$

$$x = \frac{-4 \pm \sqrt{4^2 - 4(12)(-1)}}{2(12)}$$

$$x = \frac{-4 \pm \sqrt{16 + 48}}{24}$$

$$x = \frac{-4 \pm \sqrt{64}}{24} = \frac{-4 \pm 8}{24}$$

$$x = \frac{-4 - 8}{24} \quad or \quad x = \frac{-4 + 8}{24}$$

$$x = \frac{-12}{24} \qquad x = \frac{4}{24}$$

$$x = -\frac{1}{2} \quad or \quad x = \frac{1}{6}$$

i $5x^2 - 11x + 2 = 0$

$a = 5, b = -11, c = 2$

$$x = \frac{-(-11) \pm \sqrt{(-11)^2 - 4(5)(2)}}{2(5)}$$

$$x = \frac{11 \pm \sqrt{121 - 40}}{10}$$

$$x = \frac{11 \pm \sqrt{81}}{10} = \frac{11 \pm 9}{10}$$

$$x = \frac{11 - 9}{10} \quad or \quad x = \frac{11 + 9}{10}$$

$$x = \frac{2}{10} \qquad x = \frac{20}{10}$$

$$x = \frac{1}{5} \quad or \quad x = 2$$

5 **a** $x^2 + x - 1 = 0$

$a = 1, b = 1, c = -1$

$$x = \frac{-(1) \pm \sqrt{(1)^2 - 4(1)(-1)}}{2(1)}$$

$$x = \frac{-1 \pm \sqrt{1 + 4}}{2}$$

$$x = \frac{-1 \pm \sqrt{5}}{2}$$

b $x^2 + 5x + 3 = 0$

$a = 1, b = 5, c = 3$

$$x = \frac{-5 \pm \sqrt{5^2 - 4(1)(3)}}{2(1)}$$

$$x = \frac{-5 \pm \sqrt{25 - 12}}{2}$$

$$x = \frac{-5 \pm \sqrt{13}}{2}$$

c $x^2 + 3x - 2 = 0$

$a = 1, b = 3, c = -2$

$$x = \frac{-3 \pm \sqrt{3^2 - 4(1)(-2)}}{2(1)}$$

$$x = \frac{-3 \pm \sqrt{9 + 8}}{2}$$

$$x = \frac{-3 \pm \sqrt{17}}{2}$$

d $-x^2 - 3x + 6 = 0$

$a = -1, b = -3, c = 6$

$$x = \frac{-(-3) \pm \sqrt{(-3)^2 - 4(-1)(6)}}{2(-1)}$$

$$x = \frac{3 \pm \sqrt{9 + 24}}{-2}$$

$$x = -\frac{3 \pm \sqrt{33}}{2}$$

e $-x^2 + 3x + 3 = 0$

$a = -1, b = -3, c = 6$

$$x = \frac{-3 \pm \sqrt{3^2 - 4(-1)(3)}}{2(-1)}$$

$$x = \frac{-3 \pm \sqrt{9 + 12}}{-2}$$

$$x = \frac{-3 \pm \sqrt{21}}{-2} = \frac{3 \pm \sqrt{21}}{2}$$

f $2x^2 + x - 5 = 0$

$a = 2, b = 1, c = -5$

$$x = \frac{-1 \pm \sqrt{1^2 - 4(2)(-5)}}{2(4)}$$

$$x = \frac{-1 \pm \sqrt{1 + 40}}{4}$$

$$x = \frac{-1 \pm \sqrt{41}}{4}$$

5 **g** $-2x^2 - 5x + 4 = 0$

$a = -2, b = -5, c = 4$

$$x = \frac{-(-5) \pm \sqrt{(-5)^2 - 4(-2)(4)}}{2(-2)}$$

$$x = \frac{5 \pm \sqrt{25 + 32}}{-4}$$

$$x = -\frac{5 \pm \sqrt{57}}{4}$$

h $3x^2 - 9x + 5 = 0$

$a = 3, b = -9, c = 5$

$$x = \frac{-(-9) \pm \sqrt{(-9)^2 - 4(3)(5)}}{2(2)}$$

$$x = \frac{9 \pm \sqrt{81 - 60}}{6}$$

$$x = \frac{9 \pm \sqrt{21}}{6}$$

i $3x^2 - 11x + 9 = 0$

$a = 3, b = -11, c = 9$

$$x = \frac{-(-11) \pm \sqrt{(-11)^2 - 4(3)(9)}}{2(3)}$$

$$x = \frac{11 \pm \sqrt{121 - 108}}{6}$$

$$x = \frac{11 \pm \sqrt{13}}{6}$$

6 **a** $2x^2 = 3 - 4x$

$2x^2 + 4x - 3 = 0$

$a = 2, b = 4, c = -3$

$$x = \frac{-4 \pm \sqrt{4^2 - 4(2)(-3)}}{2(2)}$$

$$x = \frac{-4 \pm \sqrt{16 + 24}}{4}$$

$$x = \frac{-4 \pm \sqrt{40}}{4} = \frac{-4 \pm 2\sqrt{10}}{4}$$

$$x = \frac{-2 \pm \sqrt{10}}{2}$$

b $3x^2 = 1 + 6x$

$3x^2 - 6x - 1 = 0$

$a = 3, b = -6, c = -1$

$$x = \frac{-(-6) \pm \sqrt{(-6)^2 - 4(3)(-1)}}{2(3)}$$

$$x = \frac{6 \pm \sqrt{36 + 12}}{6}$$

$$x = \frac{6 \pm \sqrt{48}}{6} = \frac{6 \pm 4\sqrt{3}}{6}$$

$$x = \frac{3 \pm 2\sqrt{3}}{3}$$

c $10x = 4x^2 + 3$

$4x^2 - 10x + 3 = 0$

$a = 4, b = -10, c = 3$

$$x = \frac{-(-10) \pm \sqrt{(-10)^2 - 4(4)(3)}}{2(4)}$$

$$x = \frac{10 \pm \sqrt{100 - 48}}{8}$$

$$x = \frac{10 \pm \sqrt{52}}{8} = \frac{10 \pm 2\sqrt{13}}{8}$$

$$x = \frac{5 \pm \sqrt{13}}{4}$$

d $-3 - 2x = -2x^2$

$2x^2 - 2x - 3 = 0$

$a = 2, b = -2, c = -3$

$$x = \frac{-(-2) \pm \sqrt{(-2)^2 - 4(2)(-3)}}{2(2)}$$

$$x = \frac{2 \pm \sqrt{4 + 24}}{4}$$

$$x = \frac{2 \pm \sqrt{28}}{4} = \frac{2 \pm 2\sqrt{7}}{4}$$

$$x = \frac{1 \pm \sqrt{7}}{2}$$

6 **e**

$$-7 = -4x^2 - 4x$$

$$4x^2 + 4x - 7 = 0$$

$$a = 4, b = 4, c = -7$$

$$x = \frac{-4 \pm \sqrt{4^2 - 4(4)(-7)}}{2(4)}$$

$$x = \frac{-4 \pm \sqrt{16 + 112}}{8}$$

$$x = \frac{-4 \pm \sqrt{128}}{8} = \frac{-4 \pm 8\sqrt{2}}{8}$$

$$x = \frac{-1 \pm 2\sqrt{2}}{2}$$

f

$$3x^2 = 5 + 6x$$

$$3x^2 - 6x + 5 = 0$$

$$a = 3, b = -6, c = 5$$

$$x = \frac{-(-6) \pm \sqrt{(-6)^2 - 4(3)(5)}}{2(3)}$$

$$x = \frac{6 \pm \sqrt{36 + 60}}{6}$$

$$x = \frac{6 \pm \sqrt{96}}{6} = \frac{6 \pm 4\sqrt{6}}{6}$$

$$x = \frac{3 \pm 2\sqrt{6}}{3}$$

g

$$6x = 8x^2 - 1$$

$$8x^2 - 6x - 1 = 0$$

$$a = 8, b = -6, c = -1$$

$$x = \frac{-(-6) \pm \sqrt{(-6)^2 - 4(8)(-1)}}{2(8)}$$

$$x = \frac{6 \pm \sqrt{36 + 32}}{16}$$

$$x = \frac{6 \pm \sqrt{52}}{16} = \frac{6 \pm 2\sqrt{17}}{16}$$

$$x = \frac{3 \pm \sqrt{17}}{8}$$

h

$$-2x - 3 = -6x^2$$

$$6x^2 - 2x - 3 = 0$$

$$a = 6, b = -2, c = -3$$

$$x = \frac{-(-2) \pm \sqrt{(-2)^2 - 4(6)(-3)}}{2(6)}$$

$$x = \frac{2 \pm \sqrt{4 + 72}}{12}$$

$$x = \frac{2 \pm \sqrt{76}}{12} = \frac{2 \pm 2\sqrt{19}}{12}$$

$$x = \frac{1 \pm \sqrt{19}}{6}$$

6 **i** $5x^2 - 4 = -2x$

$5x^2 + 2x - 4 = 0$

$a = 5, b = 2, c = -4$

$$x = \frac{-2 \pm \sqrt{2^2 - 4(5)(-4)}}{2(5)}$$

$$x = \frac{-2 \pm \sqrt{4 + 80}}{10}$$

$$x = \frac{-2 \pm \sqrt{84}}{10} = \frac{-2 \pm 2\sqrt{21}}{10}$$

$$x = \frac{-1 \pm \sqrt{21}}{5}$$

7 **a** $(2x + 1)(x - 3) = 3$

$2x^2 - 6x + x - 3 - 3 = 0$

$2x^2 - 5x - 6 = 0$

$a = 2, b = -5, c = -6$

$$x = \frac{-(-5) \pm \sqrt{(-5)^2 - 4(2)(-6)}}{2(2)}$$

$$x = \frac{5 \pm \sqrt{25 + 48}}{4}$$

$$x = \frac{5 \pm \sqrt{73}}{4}$$

b $(2 - x)(3x + 1) = -1$

$6x + 2 - 3x^2 - x + 1 = 0$

$-3x^2 + 5x + 3 = 0$

$a = -3, b = 5, c = 3$

$$x = \frac{-5 \pm \sqrt{5^2 - 4(-3)(3)}}{2(-3)}$$

$$x = \frac{-5 \pm \sqrt{25 + 36}}{-6}$$

$$x = \frac{5 \pm \sqrt{61}}{6}$$

7 c

$$5-2(x^2+2)=2x$$
$$5-2x^2-4-2x=0$$
$$-2x^2-2x+1=0$$
$$a=-2, b=-2, c=1$$
$$x=\frac{-(-2)\pm\sqrt{(-2)^2-4(-2)(1)}}{2(-2)}$$
$$x=\frac{2\pm\sqrt{4+8}}{-4}$$
$$x=\frac{2\pm\sqrt{12}}{-4}=\frac{2\pm2\sqrt{3}}{-4}$$
$$x=-\frac{1\pm\sqrt{3}}{2}$$

d

$$(3x-2)(x+1)=4x+7$$
$$3x^2+3x-2x-2=4x+7$$
$$3x^2+x-2-4x-7=0$$
$$3x^2-3x-9=0$$
$$x^2-x-3=0$$
$$a=1, b=-1, c=-3$$
$$x=\frac{-(-1)\pm\sqrt{(-1)^2-4(1)(-3)}}{2(1)}$$
$$x=\frac{1\pm\sqrt{1+12}}{2}$$
$$x=\frac{1\pm\sqrt{13}}{2}$$

e

$$(x+2)(2x-1)=5-x$$
$$2x^2-x+4x-2=5-x$$
$$2x^2+3x-2-5+x=0$$
$$2x^2+4x-7=0$$
$$a=2, b=4, c=-7$$
$$x=\frac{-4\pm\sqrt{4^2-4(2)(-7)}}{2(2)}$$
$$x=\frac{-4\pm\sqrt{16+56}}{4}$$
$$x=\frac{-4\pm\sqrt{72}}{4}=\frac{-4\pm6\sqrt{2}}{4}$$
$$x=\frac{-2\pm3\sqrt{2}}{2}$$

f

$$(5-2x)(x-1)=-(x+3)$$
$$5x-5-2x^2+2x=-x-3$$
$$-2x^2+7x-5+x+3=0$$
$$-2x^2+8x-2=0$$
$$x^2-4x+1=0$$
$$a=1, b=-4, c=1$$
$$x=\frac{-(-4)\pm\sqrt{(-4)^2-4(1)(1)}}{2(1)}$$
$$x=\frac{4\pm\sqrt{16-4}}{2}$$
$$x=\frac{4\pm\sqrt{12}}{2}=\frac{4\pm2\sqrt{3}}{2}$$
$$x=2\pm\sqrt{3}$$

8 **a**

$$\frac{1}{2}x^2 + 3x - 5 = 0$$

$$x^2 + 6x - 10 = 0$$

$$a = 1, b = 6, c = -10$$

$$x = \frac{-6 \pm \sqrt{6^2 - 4(1)(-10)}}{2(1)}$$

$$x = \frac{-6 \pm \sqrt{36 + 40}}{2}$$

$$x = \frac{-6 \pm \sqrt{76}}{2} = \frac{-6 \pm 2\sqrt{19}}{2}$$

$$x = -3 \pm \sqrt{19}$$

b

$$\frac{3}{4}x^2 + 2x - 1 = 0$$

$$3x^2 + 8x - 4 = 0$$

$$a = 3, b = 8, c = -4$$

$$x = \frac{-8 \pm \sqrt{8^2 - 4(3)(-4)}}{2(3)}$$

$$x = \frac{-8 \pm \sqrt{64 + 48}}{6}$$

$$x = \frac{-8 \pm \sqrt{112}}{6} = \frac{-8 \pm 4\sqrt{7}}{6}$$

$$x = \frac{-4 \pm 2\sqrt{2}}{3}$$

c

$$x^2 - \frac{1}{3}x - 1 = 0$$

$$3x^2 - x - 3 = 0$$

$$a = 3, b = -1, c = -3$$

$$x = \frac{-(-1) \pm \sqrt{(-1)^2 - 4(3)(-3)}}{2(3)}$$

$$x = \frac{1 \pm \sqrt{1 + 36}}{6}$$

$$x = \frac{1 \pm \sqrt{37}}{6}$$

d

$$3x^2 - \frac{1}{2}x - 2 = 0$$

$$6x^2 - x - 4 = 0$$

$$a = 6, b = -1, c = -4$$

$$x = \frac{-(-1) \pm \sqrt{(-1)^2 - 4(6)(-4)}}{2(6)}$$

$$x = \frac{1 \pm \sqrt{1 + 96}}{12}$$

$$x = \frac{1 \pm \sqrt{97}}{12}$$

8 **e**

$$-x^2 - \frac{2}{3}x + 5 = 0$$

$$-3x^2 - 2x + 15 = 0$$

$$a = -3, b = -2, c = 15$$

$$x = \frac{-(-2) \pm \sqrt{(-2)^2 - 4(-3)(15)}}{2(-3)}$$

$$x = \frac{2 \pm \sqrt{4 + 180}}{-6}$$

$$x = \frac{2 \pm \sqrt{184}}{2} = \frac{2 \pm 2\sqrt{46}}{-6}$$

$$x = -\frac{1 \pm \sqrt{46}}{3}$$

f

$$\frac{1}{2}x^2 + \frac{2}{3}x - \frac{5}{6} = 0$$

$$3x^2 + 4x - 5 = 0$$

$$a = 3, b = 4, c = -5$$

$$x = \frac{-4 \pm \sqrt{4^2 - 4(3)(-5)}}{2(3)}$$

$$x = \frac{-4 \pm \sqrt{16 + 60}}{6}$$

$$x = \frac{-4 \pm \sqrt{76}}{6} = \frac{-4 \pm 2\sqrt{19}}{6}$$

$$x = \frac{-2 \pm \sqrt{19}}{3}$$

g

$$5x = \frac{2x + 2}{x}$$

$$5x^2 = 2x + 2$$

$$5x^2 - 2x - 2 = 0$$

$$a = 5, b = -2, c = -2$$

$$x = \frac{-(-2) \pm \sqrt{(-2)^2 - 4(5)(-2)}}{2(5)}$$

$$x = \frac{2 \pm \sqrt{4 + 40}}{10}$$

$$x = \frac{2 \pm \sqrt{44}}{10} = \frac{2 \pm 2\sqrt{11}}{10}$$

$$x = \frac{1 \pm \sqrt{11}}{5}$$

h

$$\frac{3}{x} = 3x + 4$$

$$3 = 3x^2 + 4x$$

$$3x^2 + 4x - 3 = 0$$

$$a = 3, b = 4, c = -3$$

$$x = \frac{-4 \pm \sqrt{4^2 - 4(3)(-3)}}{2(3)}$$

$$x = \frac{-4 \pm \sqrt{52}}{6}$$

$$x = \frac{-4 \pm \sqrt{52}}{6} = \frac{-4 \pm 2\sqrt{13}}{6}$$

$$x = \frac{-2 \pm \sqrt{13}}{3}$$

8 i

$$3x = \frac{10x - 1}{2x}$$

$$6x^2 = 10x - 1$$

$$6x^2 - 10x + 1 = 0$$

$$a = 6, b = -10, c = 1$$

$$x = \frac{-(-10) \pm \sqrt{(-10)^2 - 4(6)(1)}}{2(6)}$$

$$x = \frac{10 \pm \sqrt{100 - 24}}{12}$$

$$x = \frac{10 \pm \sqrt{76}}{12} = \frac{10 \pm 2\sqrt{19}}{12}$$

$$x = \frac{5 \pm \sqrt{19}}{6}$$

Challenge Yourself

1 a

$$x^2 - kx + 8 = 0$$

$$a = 1, b = -k, c = 8$$

$$x = \frac{-(-k) \pm \sqrt{(-k)^2 - 4(1)(8)}}{2(6)}$$

$$x = \frac{k \pm \sqrt{k^2 - 32}}{2}$$

b

$$kx^2 + 5x - 7 = 0$$

$$a = k, b = 5, c = -7$$

$$x = \frac{-5 \pm \sqrt{5^2 - 4(k)(-7)}}{2(k)}$$

$$x = \frac{-5 \pm \sqrt{25 + 28k}}{2k}$$

1 c $3x^2 + 4x - k = 0$

$a = 3, b = 4, c = -k$

$$x = \frac{-4 \pm \sqrt{(-4)^2 - 4(3)(-k)}}{2(3)}$$

$$x = \frac{-4 \pm \sqrt{16 + 12k}}{6}$$

$$x = \frac{-4 \pm \sqrt{4(4 + 3k)}}{6}$$

$$x = \frac{-4 \pm 2\sqrt{4 + 3k}}{6}$$

$$x = \frac{-4 \pm 2\sqrt{4 + 3k}}{6}$$

$$x = \frac{-2 \pm \sqrt{4 + 3k}}{3}$$

d $-2x^2 - 3kx + 2k^2 = 0$

$a = -2, b = -3k, c = 2k^2$

$$x = \frac{-(-3k) \pm \sqrt{(-3k)^2 - 4(-2)(2k^2)}}{2(-2)}$$

$$x = \frac{3k \pm \sqrt{9k^2 + 16k^2}}{-4}$$

$$x = \frac{3k \pm \sqrt{25k^2}}{-4} = \frac{3k \pm 5k}{-4}$$

$$x = \frac{3k - 5k}{-4} \quad or \quad x = \frac{3k + 5k}{-4}$$

$$x = \frac{-2k}{-4} \quad or \quad x = \frac{8k}{-4}$$

$$x = \frac{k}{2} \quad or \quad x = -2k$$

2 $b^2 - 4ac = -12$

$x^2 - 4x + c = 0$

$a = 1, b = -4, c = c$

$(-4)^2 - 4(1)c = -12$

$16 - 4c = -12$

$28 = 4c$

$c = 7$

3 $b^2 - 4ac = 56$

$5 + bx - 2x^2 = 0$

$-2x^2 + bx + 5 = 0$

$a = -1, b = b, c = 5$

$b^2 - 4(-1)(5) = 56$

$b^2 + 20 = 56$

$b^2 = 36$

$b = \pm\sqrt{36}$

$b = \pm 6$

4 $3x^2 + kx + 12 = 0$

$a = 3, b = -k, c = 12$

$\Delta = 0$

$(-k)^2 - 4(3)(12) = 0$

$k^2 - 144 = 0$

$k^2 = 144$

$k = \pm\sqrt{144}$

$k = \pm 12$

5 $x^2 + (k-3)x + 3 - 2k = 0$

$a = 1, b = k - 3, c = 3 - 2k$

$\Delta = 0$

$(k-3)^2 - 4(1)(3-2k) = 0$

$(k^2 - 6k + 9) - 4(3-2k) = 0$

$k^2 - 6k + 9 - 12 + 8k = 0$

$k^2 + 2k - 3 = 0$

$(k-1)(k+3) = 0$

$k - 1 = 0 \quad or \quad k + 3 = 0$

$k = 1 \text{ or } k = -3$

Was the answer book any good?

I genuinely hope this answer book helped you learn.

And I'd love to know if it did.

So, if you don't mind, can you write a review?

Tell me what you think.

Did it help?

Were the steps clear?

Any mistakes?

Better not be.

I triple checked everything.

But seriously, let me know.

Anything.

Good or bad.

Appreciate it.

Made in United States
North Haven, CT
06 November 2026

21884526R00083